Green Energy and Technology

More information about this series at http://www.springer.com/series/8059

Samira Bagheri

Catalysis for Green Energy and Technology

 Springer

Samira Bagheri
NANOCAT
University of Malaya
Kuala Lumpur
Malaysia

ISSN 1865-3529   ISSN 1865-3537   (electronic)
Green Energy and Technology
ISBN 978-3-319-82740-7   ISBN 978-3-319-43104-8   (eBook)
DOI 10.1007/978-3-319-43104-8

Printed on acid-free paper

This Springer imprint is published by Springer Nature
The registered company is Springer International Publishing AG
The registered company address is: Gewerbestrasse 11, 6330 Cham, Switzerland

Contents

Chapter 1
Design of Catalysts, Characterization, Kinetics and Mechanisms of Reactions, Deactivation/Regeneration

1.1 Introduction

Catalysis is described as the acceleration of chemical reactions through the participation of unknown substances, known as catalysts. Catalysts usually exist as liquids or solids, but some may appear as gases. The appropriation amount of catalyst aids in prompting the thermodynamic rate of a reaction more attainable, but does not alter the composition of the thermodynamic equilibrium. In a simplified model, the reactants will react with little catalyst, hence successfully forming into the desired products at a shorter time. This is because catalysis act as an alternative route with lower reaction kinetics without disturbing the equilibrium composition of both reactants and products, thus when the catalyst is liberated the next cycle can proceed. The reaction is said to be a cyclic catalytic process. In spite of that, catalysts do not have a long-life span; by-products or structural deformation of catalyst brings to deactivation of catalyst.

1.2 Variation of Catalysis Types

Catalysis can be classified into many types such as homogeneous catalysis, heterogeneous catalysis, electrocatalysis, photocatalysis and many more. Homogeneous catalysis is discerned when both reactant and product are in the same phase, usually liquid phases. A great extend of this catalysis range of Bronsted and Lewis acids, which typically used are in metal complexes, organic acids, organometallic complexes, and carbonyls complexes. On the other hand, heterogeneous catalysis systems exist whereby the catalyst is usually in solid form and reaction occurs in either liquid or gaseous phases. Traditionally, heterogeneous catalysts are inorganic solids, but can also sometimes be organic materials for example enzymes, or organic hydroperoxides. In addition, heterogeneous catalysis

© Springer International Publishing AG 2017
S. Bagheri, *Catalysis for Green Energy and Technology*,
Green Energy and Technology, DOI 10.1007/978-3-319-43104-8_1

accounts for the great benefit comparative to homogeneous catalysis as it enhances the ease of catalyst separation from the product and are able to withstand the severity of the operating environment. Besides, a unique catalytic process entailed in oxidation or reduction by means of electron transfer is known as the electrocatalysis. Electrocatalysis is an electrochemical reaction implementing the usage of modified active catalyst electrode surface in electrolysis processes which changes the kinetics of the reaction such as in fuel cells. In photocatalysis, catalytic reaction is driven by light source whereby the catalyst absorbed light and react with reactant. In this case, both systems of heterogeneous or homogeneous can occur. Exemplary is the harnessing of semiconductor catalysts such as titanium, for degradation of organic substances by photochemical reaction. The use of substances like enzymes or microorganisms as a catalyst in various biochemical reactions to increase the rate of reaction, are known as biocatalysis. The importance of biocatalysis in factories such as the production of soft drinks is prominent for their biochemical reactions of isomerisation of glucose to fructose. In summation, the fundamental catalysis principles comprised of the reactant molecules which are coordinated to central atoms, the types of molecular species ligands or neighboring atoms at the solid matrix's surface. Even though there are disparity in the types of several catalysis, a nearer and undeniably successful participation between the distinct type of communities which represent the heterogeneous, homogeneous, and also biocatalysis must be well-foundedly supported as stated by David Parker, where at the level of molecular, there is indiscernible differences between heterogeneous and homogeneous catalysis, however there are clear differences at the level of industry (Mukhopadhyay et al. 2003).

1.3 Catalysis: A Scientific Discipline

Catalysis is firmly established scientific regime dealing not only with basic principle or mechanisms of catalytic pathways, but as well with properties, applications and preparation of various catalysts. A great number of literature scrutinized on the investigation of catalysis and processes of catalytic and the refining of development and existing of new catalysts for large scale purposes.

1.3.1 Importance of Catalysis in Industrial Application

Catalyst in chemical industry plays a crucial role in favoring the reaction and are economically beneficial as most industrial required catalytic processes. Since 1980, more than 80% of the existing industrial introduced the use of catalysts in industrial production, as well as manufacturing polymers and environmental protection products. Over 15 companies have expertise in manufacturing of thousands of catalysts pertained in many branches of industry whereby in 2008, the gross market in the world of catalysts was reckoned to be around US-$13 $\times$ 10^9.

1.3.2 A Brief Chronical of Catalysis

Earlier in 1835, a report submitted and published was first recognized by Berzelius (Alborn 1989), reviewed on numerous findings and the concept of catalysis. Nonetheless, some reactions of catalytic for instance the fermentation of alcoholic beverages from sugar or oxidation of ethanol in the production of vinegar were formulated long before. Then in the 16th and 17th centuries, catalyst is used in the production of soap by hydrolysis of fatty acids. At the same time came Mitscherlich, who was also entailed in investigating of catalytic reactions of diethyl ether from ethanol and sulfuric acid. In 1895, Ostwald (Alborn 1989; Clugston and Flemming 2000) stated the catalytic process is the acceleration of chemical reactions by the existence of unfamiliar substances that are not utilized and his underlying work was acknowledged with Nobel prize in 1909. In between year 1830 and 1900 many feasible processes were developed from flameless combustion of carbon monoxide for the catalytic oxidation process of sulphur oxides, ammonia and nitrogen oxide. In 1912, Sabatier (Alborn 1989; Che 2013) devoted to his work in saturation of alkene molecule such as of ethylene in the presence of catalyst (Ni/CO) by hydrogenation process and have received the Nobel prize for his work.

Following year, ammonia synthesis was mercenary at BASF as the Haber–Bosch (Fresco 2015) process by reinventing the development of ammonia with the presence of an iron catalyst at atmospheric nitrogen fixation. During the BASF event, Mittasch (De Jong and Geus 2000) displayed his work in synthesizing ammonia with iron catalysts. In 1938, production of high-pressure and temperature catalytic hydrogenation for use of synthetic fuel in the presence of an Fe catalyst was done by Bergius (Alborn 1989; Stranges 1984). Besides, others catalytic industrial processes were the methanol production in the presence of zinc oxide complex and the fractionation of petroleum into gasoline by acid-activated clay catalyst, as shown by Houdry (Alborn 1989; Busca 2007). In 1932, Ipatieff et al. (Alborn 1989; Olah and Molnar 2003) reported the first addition reaction of isobutane to carbon olefins (C3–C4) over $AlCl_3$, which leads to a branched alkyl chain (C7–C8). His innovation in the field of petroleum chemistry has endowed owing to UOP (USA) as commercial process. Alongside, in 1925, Fischer and Tropsch (Smil 2004) discovered a commercialized renowned reaction by converting gas to liquids in presence of alkalized iron catalyst. Before World War II, fabricating aliphatic aldehydes to olefins in the presence of cobalt carbonyls (hydroformylation process) was the acme in the catalytic industry in German, discovered by Roelen (Sturgeon et al. 2014). His work was further scrutinized by Ruhr-Chemie in 1942 and was commercialized and known as Oxo Synthesis. During and after World War II (until 1970), several reactions involving catalyst were being discerned on an industrial scale.

1.3.3 Principle of Active Sites of Catalyst

Active site is a site on the catalyst surface that enable the formation of strong chemical bonds with the reactant whereby the sites (unsaturated atoms in the solid form) are resulting from surface defects. In accordance with Sabatier's principle of catalyst active sites, stated the binding energy of the reactants to the catalyst must be strong enough to produce reaction intermediates, but has to be weak enough to allow dissociation of catalyst surface of the products in order for more reactions to take place at that site. Besides, Langmuir adsorption isotherm catalytic model stated his ideology, where an assemblage of sites which were energetically indistinguishable and non-interacting, would adsorb just one molecule from the gas phase in an active site (van Santen 2017). However, he was already aware that the hypothesis of indistinguishable and non-interacting sites was a guess, which would not hold for the real surfaces. Thus, for the first time, he developed the surface science approach to heterogeneous catalysis whereby the adsorption and desorption pressure are in equilibrium.

The different phases of active sites on solid catalyst surfaces and its repercussion were highlighted by Taylor (Thomas et al. 2005). The differences of coordination numbers of surface atoms will govern to distinct reactivities and activities of the relative sites. The statement was in agreement with Schwab's delineation theory (Ramírez et al. 2004), which stated that line defects consisting of atomic steps are of utmost importance and was later affirmed by the study of surface science on stepped mono-crystal metal surfaces (Xuereb and Raja 2011). Moreover, different coordination numbers of surface atoms in a single component solids show distinct surface composition and dissimilar for each crystallographic plane in various component materials (Eckert and Wachs 1989). Then, Boudart and co-workers (Goodman 1996) invented the term structure-sensitive/insensitive reactions based on precise kinetic measurements and on the Taylor's principle of inequivalent active sites. His reaction is one in which all sites appear to display the same activity on several planes of a single crystal. This is in agreement with Taylor's theory whereby the surface of a catalyst readjusted itself in the feed with minimal surface free energy, which are attainable by surface-reconstruction (Roberts et al. 2015; He 2013), and characterization of active sites are essential under working conditions of the catalytic system.

The fundamental of active sites is not limited to only metals but also include ions, Lewis and Brønsted acids, immobilized enzymes, and organometallic compounds. Formation of multiplets (Price et al. 2000) or ensembles (Thomas et al. 2005) in more than one species (or atom) are available in active sites. A compulsory prerequisite of activation of the active sites depends on the availability of absorbing materials by chemical binds from the fluid phase. Therefore, available coordination sites are necessary. According to Burwell and co-workers, (Clarke and Creaner 1981; Tanaka et al. 2012) the term coordinatively unsaturated sites in correlation with homogeneous organometallic catalysts define the active sites as a group of atoms which are implanted in the matrix surface whereby the neighboring atoms function as ligands. The entity and ligand effects are further discussed (Zhang et al. 1991) and quantum chemical treatments of geometric composite and electronic ligand effects on metal alloy surfaces are studied in detail (Sachtler 1984).

1.3.4   Catalysis Surface Coordination Chemistry

The assemble of intricate surface by atoms or molecules are perceived to usually to be similar of a general structure likewise to molecular coordination complexes. The affix of molecules in surface complexes can be illustrated with a localized view (Mayer and Hafner 2011; Chiesa et al. 2010). Therefore, significant event materialized on the surface of solid catalysts may be recounted in the backbone skeleton of surface coordination chemistry or surface organometallic chemistry (Korlach and Turner 2007). However, this is in contrast to the band theory of catalysis, which aim to associate the catalytic performance with electronic properties (Coq and Figueras 2001) as studied by Stone in his theory of limitation in oxide catalysis (Stone 1975).

1.4   Catalysis Modifiers and Promoters

The feasibility, operations of industrial catalysts are commonly enhanced by modifiers or additives (Hutchings 2001; Mallat and Baiker 2000). A modifier is also known as a promoter when it aids in the catalytic activities in terms of reaction rate per site. Nonetheless, the addition of modifiers may also affect the catalytic performances in an unfavorable manner whereby it can be poisonous. However, the eminence between promoters and poisons is not complicated for reactions in parallel or consecutive steps producing several products, of which only one is the desired product in which presence of high selectivity is essential. Thus, by adding poisonous substances, the selectivity can be enhanced by prompting favorable reactions. In exothermic reactions, extreme high reaction rates may promote to an uncommon rising in temperature, which harness in unfavorable products for instance CO and CO_2 in a selective catalytic oxidation. Besides, insubstantial catalyst stability may also take place which leads to degradation of the catalyst. As a result, a modifier is needed in a reaction which lower the reaction rate so that the steady-state temperature and reaction rate can be sustained. In addition, modifiers enable the binding energy of an active site or its structure to be altered, or destroy the composition of atoms complex.

A molecular technique toward the apprehension of promotion in heterogeneous catalysis was introduced by Hutchings. For example, the synthesis of Ferrum-based ammonia catalyst is aided by alumina and potassium oxide. Alumina appears as a promoter, as it impedes the sintering process of pure iron metal whilst stabilize most active sites on the iron surface (structural promoter). On the other hand, K_2O materialized in affecting the adsorption kinetics and splitting of N_2 as well as the binding energy of N atoms on adjacent ferrum sites (electronic promoter). Another examplary is the introduction of Co to MoS_2-based catalysts assisted on transitional aluminium oxide have a promising outcome on the rate of hydrodesulfurization of sulfur complexes at Co/(Co + Mo) ratios below *ca.* 0.3. The active phase is suggested to be the CoMoS phase, which comprises of anchored edges of MoS_2 platelets on Co atoms.

1.5 Catalyst Active Phase—Support Interactions

Numerous valuable opinions proved that in elucidating event that is relevant to obvious catalysts classes. The phase of active usually compromised of oxide, metal, and sulfide undergoes interactions of active phase-support in the supported catalyst system (Bartholomew 2001). The interaction between the materials of active phase and support are dependently determined via the Young's equation-surface free energies and the interfacial free energy between both components (Bartholomew and Farrauto 2011). For example, when compared to the material of typical oxidic support, for example TiO_2, V_2O_5 active transition metal oxides have surface free energies which are relatively low, thus interactions between both the support and materials materialized with the exception of transition metal oxides of SiO_2-supported are favorable. In consequences, when the treatment with thermal to the mixtures of oxide is at relatively high temperature (above the Tammann temperature) promoting the mobility of the active oxide. In corresponding, the spreading and saturation of the active transition metal oxides can occur on the surface of support and the catalyst type of monolayer will be formed. Besides, usually, the surface free energies of pure metals and transition were high (Raybaud 2007), thus, an agglomeration of small particles or crystallites leaning to surface area reduction is favorable and stabilization can be done by deposition of surface supports on the metal-support interactions (MSI). The properties of the catalysis reaction of that metal particle were greatly influenced by the nature of the material of support.

Besides, reinstation of dispersion of metal with high degree, can be reduced under sufficiently mild conditions. The number of metal atoms that was exposed at the surface of the particle to the total number of metal atoms in the particle ratios was defined as dispersion, D. Experimentally, the capacity of adsorption for hydrogen gas and carbon monoxide are tremendously plummeting and the oxide support is slightly decreasing when the precursor on the support for catalytically active metal were decreases in hydrogen gas condition at the temperatures more than 770 K (Moulijn et al. 2001). This is due to the particle of metal which has been encapsulated by material of support oxide. On the other hand, effect of electronic properties of the particle of metal was influenced by the oxide of support in the state of SMSI, where declining of the capacity of adsorption seems to be predominantly because of the effect of geometric, hence results in the solitary of the surface of metal. Calcination at low-temperature may bring about a overlayers of oxide which are well-dispersed, while at low temperature reoxidation with direct reduction will result in particles of metal which are highly dispersed. In addition, at high calcination temperature, assemblation of surface compound may likely happen through a reaction of a solid-state between the support and also the precursor of active metal. However, reduction in temperatures, which are high leads to agglomeration of particles due to strong cohesive forces. Therefore, in either condition, the cohesive interaction between metal particles must be in a mobile state in order to stabilize the system. Alternatively, if adhesive forces are over powering encapsulation are likewise to happen on the reduced surface area.

1.6   Abstraction of Catalyst Shape-Selectivity

The selectivity of the shape of zeolites and also its corresponding materials have been explored due to their crystalline structure and the diameters regular micropores, which are ascertained by the structural material. The geometric limitation may behave on the products desorption, reactants sorption and also catalyzed reaction transition state. Relatively, the effects of the selectivity of shape already denoted make available for use as selectivity of reactant shape, hampering the selectivity for shape of the transition state, and selectivity of the product shape (De Vos et al. 1994). Examplary, when compared to 1-methylhexane, n-heptane has a smaller diameter of kinetic. Therefore, the micropores was not able to be entered by 1-methylhexane, so the n-heptane shape selective cracking will occur when presented in the system. Another example of shape selective control is bimolecular reaction and the 1,2,4-trimethylbenzene will be formed. A transition state of 1,3,5-trimethylbenzene are bulkier than 1,2,4-trimethylbenzene, thus the product will not be materialized if the geometry and the size of the pore is being altered to the required transition state.

1.6.1   The Fundamental of Catalytic Cycle

The most basic catalysis principle in catalytic cycle reaction, are being redefined by Boudart (Kozuch and Martin 2012): "Substance that will convert reactants to products, by the cycle of repeated elementary steps and interrupted is defined as a catalyst". The active site of catalytic can be unavailable initially, however, it can be activated in the reaction of catalytic. The cycle should not be disturbed and in continuous repetition in order for it to be catalytic. Besides, the intermediates which are reactive can be approached through the steady-state of kinetic quasi whereby the number of turnovers must be greater than unity as the catalyst span is determined by the cycle numbers right before it perish. The total amount of catalyst added is rather less compared to the amounts of products and reactants that were involved.

1.6.2 Reaction Kinetics of Heterogeneous Catalytic Cycle

The catalyst fundamental of catalytic action is its catalytic cycle. The mechanism of catalytic reaction will be explained through the succession of simple steps of the reaction of the cycle, involving diffusion of surface, adsorption, desorption for attaining the reaction of kinetic and also the adsorbed species chemical transformations. An early study has been published on the kinetic reaction that was catalyzed heterogeneously (Boudart 1985). In the system of macroscopic, the rate reaction for catalytic is fashioned through suitable equations of empirical to data of experiment, in order to describe its dependence of rate constants exponentially on temperature and to determine its concentration and pressure dependence. Deduction about kinetic models contribute to the relationship between the external variables and also the intermediates surface coverages, a technique that led to the equation of Temkin (Borodziński and Cybulski 2000) modeling the kinetics synthesis of ammonia. Ameliorate models of kinetic can be enhanced by making available of processes of atomic on surfaces and also the surface characterization and also identification. Then, the reaction progress was depicted via a strategy of microkinetics by fashioning the kinetics of macroscopic via association of parameters of macroscopic and processes of atoms in the structure of an apt model of continuum linking to external variable such as temperatures and also partial pressures through the model of Langmuir of a surface lattice comprising of almost alike noninteracting sites of adsorption.

1.6.3 Generation of Solid Catalysts

The generation of a process of catalytic encompasses the catalyst findings and also a reactor which is suitable, as well as occurrence in a steps succession at different levels. In this process, reactors for small-scale function as a screener to identify the formulation of optimal catalyst. Since the screening of sequential and also the catalyst generation are tedious and costly, techniques of high throughput experimentation (HTE) (Senkan 2001) which allows testing of the parallel of catalyst at small amounts in the systems which are automated have raised tremendous interest.

1.7 Solid Catalysts Classification

In processes that are large scale, the solid catalysts are significantly vital (Clark 2002) for the fuels, chemicals and also pollutants conversion. However, common industrial catalysts composed of numerous components and phases. This complication generally makes it harder to evaluate the structure of the catalytic material. The variation of families of existing catalysts is described; such as (1) unsupported catalysts; (2) supported catalysts; (3) confined catalysts; (4) hybrid catalysts; (5) polymerization catalysts, and etc. The selected examples not only inclusive of materials which are used in the industry, as well as materials which are well developed for technological application.

1.7.1 Unsupported Catalysts

Important classes of chemical compounds comprising of oxygen are oxides, in which the O atom are strongly electronegative compound. Metals oxides of usually solids and their unsupported properties vastly depend on the bonding character between metal and oxygen. Metal oxides have large variation of electronic properties which composed of metallic conductors, insulators, semiconductors, high-temperature superconductors and superconductors. Metal oxides concoct a massive and major class of catalytically active materials, surface properties and chemistry being governed by their structure and composition, the chemical bonding, and the coordination of surface atoms and hydroxyl groups on the crystallographic faces. Oxides of metal can have non-complex composition, like binary oxides, but numerous technologically important oxide catalysts are complex multi-component materials.

1.7.1.1 Simple Binary Oxides of Catalyst

Simple binary oxides of basic metals act as solid acids, bases or amphoteric materials (Kreuer 1997). These properties are associated with their dissolution behavior in contact with aqueous solutions. Amphoteric oxides can form both cations and anions in acidic and basic medium. Acidic oxides dissolve with the formation of acids or anions. Transition metal oxides in their highest oxidation state behave similarly. Basic oxides form hydroxides or dissolve by forming cations or bases. The dissolution behavior must be taken into account when distinct oxides are used as supports and permeate into aqueous solutions of the active phase precursor (Valigi et al. 1999).

Aluminas such as bayerite, boehmite, and gibbsite are amphoteric oxides, which can be in the form of different phases depending on the nature of the precursor and the thermal decomposition. The structures of aluminas can be illustrated as a closely packed layer of oxo anions with Al^{3+} cations assemble evenly in vacancy positions between tetrahedral and octahedral. Stacking variations of the oxo anions displayed in the various crystallographic forms of alumina. The most commonly used transitional phases are h- and g-Al_2O_3, which are known as the defect spinel structures (Paulik et al. 2007) in presence of Al^{3+} cations in both tetrahedral and octahedral sites. The existing hydroxyl groups can be doubly or triply bridging with the inclusion of Al^{3+} in tetrahedral/octahedral positions or be terminal. The behaviour of the OH species ranges from very weakly Brønsted acidic to rather strongly basic and nucleophilic (Ryczkowski 2001) thus, a pronounced rich surface chemistry and specific catalytic properties of alumina was attained. The particle size and surface area of these materials can be governed by the parameters, such as their thermal stability and redox; where high thermal stability and long life span of catalyst enhanced the activities of the supported active phases (Auroux and Gervasini 2003).

Silicas structures like quartz and cristobalite are a weakly Brønsted acidic oxides (Rimola et al. 2013). Silica is typically used in catalysis as amorphous silica in which the framework of silica is associated with SiO_4 tetrahedral, with each O atom bridging two Si atoms by covalent bonds. At the fully hydrated surface, the unsupported structure is discontinued by hydroxyl silanol groups (Skjærvø 2016) during calcination at temperatures below 473 K, which may be exposed with germinal groups $Si(OH)_2$ (Kemball and Dowden 1981). Besides, silicas are able to form hydrogen-bonding (Nakamura and Matsui 1995) due to the weakly Brønsted acidic surface of hydroxyl groups in enhancing any catalytic activity. Thus, due to this reason silicas can not be used as active catalysts, but they also play a significant role as oxide supports and fabrication of functionalized oxide supports. Man-made silicas can be prepared by governing the parameters such as porosity, surface area, particle size and morphology, and mechanical stability (Peyratout and Daehne 2004). Moreover, the crystalline microporous silica can be prepared by hydrothermal synthesis (Borm et al. 2006) with corresponding MFI structure and large-pore mesoporous structures (Xue and Ma 2013).

Magnesium oxide is a basic solid with octahedral coordination of magnesium and oxygen. Calculations of molecular orbital depicted that the electronic structure are highly ionic in presence of $Mg^{2+}O^{2-}$ ions being an ideal model of both surface structures and bulk (Tossell 1975). This representation seems to be physically accurate for MgO smoke, which may be known as a prototype support of crystalline metal oxide (Burke et al. 2015). Even though the (100) plane is electrically non-charging, hydroxyl groups are present on the surfaces of polycrystalline MgO and association with O^{2-} ions resulted in its basic properties and weak Lewis acid sites on the coordinative unsaturated Mg2þ ions. These characteristic influences the surface chemistry of MgO where chemisorbed separate to create a surface boundary between carbanions and surface hydroxyl groups (Lundie et al. 2005). Ion of $Mg^{2+}O^{2-}$ pairs with each other in a 3- or 4-coordination (low coordination) appears to play an important role on the MgO surface after activation at high temperatures has been portrayed and the distinct reactivity of cubic centers has been analyzed (Zhidomirov and Chuvylkin 1986).

Transition metal oxides (Cotton et al. 1988) are structurally denoted as more or less dense packings of oxide anions and the interstitial sites are occupied with cations. The bonding of the transition metal oxides is a mixed ionic-covalent, with some metallic character. The surface of these materials is usually partially occupied by OH^- groups, so they are somewhat acidic in behavior. Characteristic of redox catalytic reaction are reflected by the differences in oxidation states and the chances of producing mixed-valence and nonstoichiometric compounds. The typically utilized transition metal oxides are mostly from the early transition metals and application of these materials are particularly in fields of selective oxidation and dehydrogenation reactions.

Titania TiO_2 can appear in two forms of crystallographic state that is anatase and rutile. Anatase are commonly altered as it generates a wide surface area, even though it is a metastable phase and may undergo slow transfiguration into a more stable (thermodynamically) rutile phase above ca. 900 K. In this system, vanadium impurities aids in accelerating the rutilization above 820 K while impurities such as surface sulfate and phosphate appears to stabilize the anatase phase. The anatase to rutile phase transition acts as an important role in selective oxidation and NOx reduction catalysis thus, must be sensitively controlled for supported VOx/TiO_2. Titania has a wide band gap and are a crucial material for photocatalysis with a promising semiconductor properties (Pelaez et al. 2012).

1.7.1.2 Catalyst Complex of Multicomponent Oxides

Aluminum phosphates are compounds whose structures are alike with zeolites (Walawalkar et al. 1999). It can be reckoned as zeolites in which the T atoms are silicon and aluminium. Recently, aluminium phosphates have been known as zeotypes, the T atoms consist of aluminium and phosphorus. Aluminium phosphates framework generally is composed of Al/P atomic ratio of 1/1, so that the material is in neutral. Thus, the material is non-acidic and are lessly seen as usage in any application as catalysts. Nonetheless, in these cases, introduction of acids by substituting Al^{3+} enables production of metal aluminophosphates for example, partial substitution of formally pentavalent phosphorous by Si^{4+} to yield silicoaluminophosphates with a wider range stretch of pore diameters.

Clays for examples montmorillonite are aluminosilicate minerals. Montmorillonite is an aluminohydroxysilicate with a ratio of 2:1 clay constituent whereby one octahedral AlO_6 layer is sandwiched between two tetrahedral SiO_4 layers. Montmorillonites having ion-exchange capacity and reversibly swellable can be utilized as a catalyst supports. Mixed metal oxides pillars are catalytically active multimetal phase oxides which generally consist of one or more transition metal oxide, hence, displaying important chemical and structural complexity (Govindasamy et al. 2010).

Vanadium phosphates are the starting material for the VPO catalysts, which catalyze ammoxidation reactions and the selective oxidation of n-butane to maleic anhydride. It is suggested that the catalytic properties of the VPO system depend on the crystalline vanadyl pyrophosphate phase $(VO)_2P_2O_7$. The starting material of vanadium phosphate undergoes changes in oxidizing and reducing conditions.

Bismuth molybdates are one of the fundamental catalysts for selective oxidation and ammoxidation of hydrocarbons (Grasselli et al. 2005). The catalytically chief phases situated in the compositional range Bi/Mo atomic ratio between 2/3 and 2/1 and are $g\text{-}Bi_2MoO_6$, $b\text{-}Bi_2Mo_2O_9$, and $a\text{-}Bi_2Mo_3O_{12}$. Application in industrial scale of Bi molybdate catalyst was optimized in numerous steps where the material is bear on 50% SiO_2 and was further optimized resulted a catalyst with the empirical formula (K, Cs) (Ni, Mg, Mn)$_{7.5}$(Fe, Cr)$_{2.3}$Bi$_{0.5}$Mo$_{12}$O$_x$.

Scheelites are multicomponent oxides espouse the scheelite, $CaWO_4$ structure. This compound withstands replacements of cation regardless of valency allot that A is a larger cation than B and that there is net charge. Besides, scheelite structure is usually stable with 30% or more vacancies in the A cation sublattice.

Perovskite is a mineral with component of $CaTiO_3$. The typical characteristics are it is highly charged, similar to that of the scheelite-type oxides, B cation and a large cation A, often having a low charge and able to withstand a large variation of compositions.

Heteropolyanions are polymeric oxo anions (polyoxometalates) fashioned by condensation of more than two kinds of oxo anions. The amphoteric metals of Groups 5 and 6 in the +5 and +6 oxidation states, respectively, create weak acids which promptly condense to form anions containing molecules of the acid anhydride. Process of condensation can also materialize with other acids to produce heteropolyacids and salts.

1.7.1.3 Metals and Metal Alloys

Metals and metal alloys are used in large amount, unsupported catalysts in only some studies. Metal grids or gauzes are utilized as unsupported catalysts in strongly exothermic reactions which involves catalyst beds of small height. For instance, dehydrogenation of methane to formaldehyde in ammonia oxidation utilises platinum–rhodium grids.

Skeletal catalysts (Raney-type), are materials which are particularly utilized in the reduction of saturated organic compound by catalyst (hydrogenation reactions). Spite of that, their utilization is constricted to liquid-phase reactions. They are applied specifically in the production of pharmaceuticals and fine chemicals. Skeletal catalysts are concocted by the specific removal of aluminum from Ni–Al alloy particles by percolate with aqueous NaOH. Among the benefits of skeletal metal catalysts is that they can be kept in the active metal form and no pre-reduction is required.

Fused catalysts function as an alloy catalyst. The fabrication from a uniform melt by fast cooling may produce a metastable material with attainable compositions. Typical fused catalyst is amorphous metal alloys (metallic glasses) (Kear et al. 1979) and oxide materials. Oxides displayed an intricate and active internal interface structure which is practical for either direct catalytic application in oxidation reactions or in predetermining the micromorphology of resulting catalytic materials. The framework of such catalyst is the numerous proselytized iron oxide starting material of catalysts used for ammonia synthesis (Lazcano and Peretó 2010).

1.7.1.4 Carbides and Nitrides

Single metallic carbides and nitrides of transition metals, usually espouse non-complicated crystal structures which are arrayed in cubic close-packed (ccp), hexagonal close packed (hcp), or simple hexagonal (hex) arrays. Carbon and nitrogen atoms filled in the interstices positions between the metal atoms. This enhance the distinct materials properties in terms of melting point, hardness, and strength on the other hand, their physical characteristic displayed those of ceramic materials, with metals-alike electronic and magnetic properties. carbides of carbon give electrons to the d orbital of the metal, therefore favoring the electronic characteristics.

1.7.1.5 Carbons

Even though carbons are commonly applied as a catalyst supports, utilization as catalysts on its own were applicable (Zhu et al. 2000). Implementation of catalytic carbons in industrial scale involves the selective oxidation of hydrogen sulfide to sulfur in the presence of oxygen in the gas phase, the oxidation of sulfurous to sulfuric acid, and the reaction between phosgene and formaldehyde.

1.7.1.6 Composition of Metal–Organic

Composition of metal–organic are highly permeable and containing three-dimensional network crystalline solids. MOFs are homologous to zeolites where the components are arranged systematically of the structural units which increases to a system of channels and cavities on the nanometer length scale. A highlight for the progress of MOFs was the fabrication of MOF^{-5} whereby the material comprises of tetrahedral $Zn_4O_6^+$ clusters held by terephthalate groups. Thus, by apt selection of the linker length, the size of the promising pores can be fashioned.

1.7.2 Supported Catalysts

Supported catalysts play a significant part in providing high surface area and stabilizes the dispersion of the active component in many industrial processes. Interactions of active phase support are governed by the surface chemistry of the support which are accountable for the dispersion and the chemical state of the catalyst. Supports may actively involved with the catalytic process. To attain the highly stabilize disperse active phase and high surface areas, supports are typically porous materials having stability at high temperature.

1.7.2.1 Supports

Many of the unsupported materials are commonly binary oxides and ternary oxides. Others probable catalyst supports are aluminophosphates, kieselguhr, calcium aluminate and bauxite. Carbons such as charcoal or activated carbon can also be used as supports unless oxygen is needed at high temperatures in the system.

1.7.2.2 Coated Catalysts

Coated catalysts are catalytically active layers spread on the inert structured surfaces. The active layers comprise of either unsupported or supported catalysts. To date, the utilization of coated catalysts has become enormously popular. Examplary systems are, eggshell catalysts deposited on an inert carrier, applications of monolithic honeycombs or for multiphase reactions for environmental studies, structured packings and catalytic-wall reactors and filters for flue gas treatment and diesel exhaust after-treatment. Benefits of coated catalysts are highly efficacious mass transfer, superb usage of the active mass, low pressure drop and high selectivity at low diffusion lengths.

1.8   Fashioning of Catalysts

Designation of catalysts in industrial scale can be subdivided into two categories basically

1. unsupported catalysts
2. Supported catalysts

1.8.1   Unsupported Catalysts

Preparation techniques involved in the commercial production of unsupported catalyst are mechanical treatment, fusion of catalyst components, precipitation and coprecipitation, flame hydrolysis, and hydrothermal synthesis (Campanati et al. 2003; Ertl et al. 2008).

Mechanical treatment of catalytic active materials with structure stabilizers, promoters, or pore-forming agents, is among the effortless preparation techniques. A recent enhancement in the efficiency of numerous aggregates for the mechanical treatment of solids resulted in positive feedback in its activities. Besides, the main advantage of these techniques is that production of wastewater is avoided.

Fusion of components or precursors is designed for the development of a small group of unsupported catalysts. The fusion process allows the fabrication of alloys containing of elements which is insoluble in solution or present in the solid state. Nonetheless, characterizing of unsupported catalysts by fusion process is costly and energy-consuming.

The most commonly used techniques for the preparation of unsupported or supports catalysts are *precipitation and coprecipitation*. In spite of that, both techniques have the major drawback of creating large volumes of salt-containing solutions in the precipitation and in rinsing stage. The most common sources are basically metal salts, however, formates, acetates, and oxalates are used in some cases. In industrial scale, nitrates or sulfates are favored.

Sol-gel synthesis participates in the formation of a sol, followed by the production of gel. Process of hydrolysis and partial condensation of a salt or a metal alkoxide formed a sol. Then, continuation of condensation process results in a three-dimensional network in the formation of a gel. Exposed parameters of its sol-gel influence the porosity and the strength of the gel. The latter show the growth of crystalline fraction of the gel in elevated temperature, slow coagulation, or hydrothermal post-treatment conditions.

The flame Hydrolysis process is a concoction of the catalyst or support starting material in the presence of hydrogen and air. The material is fed into constant flame in the operating reactor. Typically, the starting material undergoes chemical breakdown by steam/water to produce oxides by hydrolysis (Campanati et al. 2003).

Thermal decomposition of the starting material of inorganic or organic metal catalyst is occasionally used in production of industrial catalyst. As example, mixtures of Cu- and Zn $(NH_3)_4(HCO_3)_2$ disintegrate at 370 K and are transformed into binary Cu–Zn carbonates during calcination when low temperature water gas shift catalysts were used.

Hydrothermal process is an intricate technique comprising of three fundamental stages which is consummation of supersaturation, crystal growth, and nucleation. It is influenced by the time, temperature, alkalinity and hydrogel molar composition.

1.8.2 Supported Catalysts

The wide application in industrial catalysis of supported catalysts influences in the evolution of many apt preparation methods on an industrial scale. These techniques are similar with those used in the production of unsupported catalysts, for instance, precipitation, thermal decomposition of metal—inorganic/organic complexes and mechanical treatment.

Mechanical treatment is utilized in the molding of a catalyst starting material with a support (e.g., production of kieselguhr-supported Ni).

Impregnation by filling the pore of a carrier with an active phase is commonly utilized in production techniques for supported catalysts (Nijhuis et al. 2001). The technique applies pores filling of the support with a solution of the catalyst precursor to attain the desired loading. If higher loadings with active phases are essential, repetition the impregnation of the intermediate after drying or calcination.

Adsorption is a method to attain consistent deposition of little amounts of active component on a support. Particles or powders adsorb equilibrium quantities of exposed metal salt ions, in agreement with the adsorption isotherms. Adsorption is either cationic or anionic, relying on the properties of the carrier surface.

Ion exchange takes part in the exchange of ions besides protons. Lower valence ions, such as Na^+ can be interchanged with higher valence ions, for example, Ni^{2+}. This technique basically used in the preparation of metal-containing zeolites in petroleum-refining processes.

Thermal decomposition techniques utilized the preparation of unsupported inorganic or organic complexes catalysts by thermal decomposition of precursors in the presence of a support. Supports are typically in powder or preshaped form.

Precipitation onto the support is prepared in the case of unsupported catalysts. Powders which are slurried in the salt solution in addition of alkali are usually used as a support. Fast mixing is necessary to eschew precipitation in the bulk and usage of alkalis (urea) ensured constant precipitation.

Reductive deposition is a preparation technique in which novel metals are suspended on the carrier surface by reduction of aqueous metal salts, mainly chlorides or nitrates, with hydrogen gas, formaldehyde and hydrazine as an agent.

Heterogenization of homogeneous catalysts is the attachment of metal complexes to the surface or entrapment in the pores of the inorganic or organic support (De Vos et al. 2002). The catalysts used are generally in selective hydrogenations in the production of pharmaceuticals or fine chemicals. In rare cases, heterogenized enzymes can also be done. Industrial application is the isomerization of glucose to fructose in the production of soft drinks.

1.9   Characterization of Catalyst

Parameters such as surface chemical composition, morphology and texture, structure of solid catalysts and phase composition greatly influences the catalytic activity. Hence, various physical and chemical techniques are utilized in research of catalysis to characterize the compound to determine the interaction between structure and performance of catalysts. These techniques involve traditional procedures as well as the latest techniques for the study of the surface chemistry and its physical state (Dubois and Nuzzo 1992; West 2007).

1.9.1   Physical Properties

1.9.1.1   Specific Surface Area and Porosity

The specific surface area of a catalyst is dictated by measuring the volume of gas (nitrogen gas), to allow a single molecular layer following to the

Brunauer–Emmett–Teller (BET) technique. This technique uses the physisorption of the test gas to deduce the monolayer capacity. From this approach, the volume adsorbed at a given equilibrium pressure can be measured by static or dynamic methods. Calculation by the addition of the internal and external surface areas gives the total surface area of a porous material where classification of pores according to IUPAC definitions are as followed; (1) micropores (pore width <2 nm), (2) mesopores (pore width 2–50 nm), and (3) macropores (pore width >50 nm).

1.9.1.2 Size of Particles and Dispersion

The size of the metal particles disseminates on a support deserves precise consideration as the catalytic properties of supported metal catalysts are influenced by the metal surface area and particle size. The metal dispersion D is given by D ¼ NE/M$_T$, where NS is the number of metal atoms exposed at the surface and M$_T$ is the total number of metal atoms in a given amount of catalyst. The fraction of surface atoms D can be analyzed if the number of metal atoms exposed at the surface is available in the studies and can be deducted by chemisorption measurements with adsorptives that bind to the metal strongly with the support at the specific temperatures and pressures. Maximum extend of values of the chemisorbed amounts allows number of metal atoms exposed at the surface to be calculated if the chemisorption stoichiometries are known. Dispersion is straightwordly analogous to particle size distribution and particle size. Determining the mean size of crystallite particles distributions can be characterized from X-ray diffraction line broadening, and small-angle X-ray scattering (SAXS). Approach of electron microscopy enables the observation of catalyst morphologies of relevant range of particle sizes. The structural analysis can be obtained from lattice imaging and electron-diffraction methods.

1.9.1.3 Structural and Morphology

X-ray powder diffraction (**XRD**) is a method based on the differentiation of the observed set of reflections on the catalyst sample with those of uncontaminated reference phases for the distinguishing of present catalyst phases. In XRD studies, in situ can be prepared out on the working catalyst, and the use of synchrotron radiation enables dynamic experiments in real time (Gai 1999).

X-ray Absorption Spectroscopy (XAS), is a technique where the utilization of XRD for structure analyses seems to be possible due to their high photon flux. The synchrotron facilities are the preferred elements for XAS experiments as the ejection of a photoelectron from a main level of an atom by absorption of an X-ray photon.

In Electron Microscopy and Diffraction, when electrons perforate through the sample in an electron microscope, differences are formed by differential absorption or by diffraction phenomena. Electron micrographs of catalyst materials contribute to identification of phases, elemental distributions and compositions, and their morphologies.

High-resolution electron microscopy (HREM) can be done in conventional transmission electron microscopy (CTEM) instruments by changing the imaging mode, or at electron energies of 0.5–1.0 meV. HREM images can be directly related to the atomic structure of the material where lattice fringes and the tenacity of the spacings of atomic planes can be solved. Identification of support particles as well as the crystal structure of metal particles in the range lower to 1 nm can be investigated.

Scanning electron microscopy (SEM) is a useful tool in analyzing samples topographies. SEM produces images by scanning the focused probe beam subtly in a parallel scanning line pattern across the specimen surface. Emitted signals, for instance, secondary electrons, are detected and employ for image formation. The best resolutions that can be attained with current generation SEM instruments are roughly 1 nm.

Selected area electron diffraction (SAED) analyses on phase compositions and structures at a microscopic level. The integration of microdiffraction patterns and brightfield images allows information of exposed facets and shapes of solid catalysts in dispersed phases.

Analytical electron microscopy (AEM) enables the characterization of the elemental composition of a solid catalyst at the microscopic level of energy scattering of the electron induced X-ray emission. Energy scattering spectroscopy (EDS) is sensitive for elements with atomic numbers $Z > 11$ (Geiger 2009).

1.9.2 Chemical Properties

1.9.2.1 Catalyst Surface Chemical Composition

In catalytic properties, atomic framework of a catalyst surface can be determined by electroscope which gives details on the atomic composition within the top most atomic layers such as the number of atomic layers providing to the measured signal. Profiles of the concentration can be acquired by ion bombardment of the surface in ultrahigh-vacuum (UHV) conditions. The principal for the identification of atoms on surfaces of solid materials by electron spectroscopies, can be characterized by Auger electron spectroscopy (AES) and X-ray photoelectron spectroscopy (XPS) electronic binding energies. As for ion spectroscopies, slow-energy ion scattering (LEIS) and Rutherford backscattering (RDS), provides surface atoms identification by their nuclear masses. Ion bombardment of surfaces is equipped by sputtering processes (surface etching) which aid in removal of secondary ionic and neutral particles. The compound is then characterized by mass spectroscopic methods to obtain additional information on the surface properties of catalyst.

1.9.2.2   Acidity and Basicity

Acid and base properties, can be characterized by controlling the acid or base strength, the number of sites per unit surface area of a solid catalyst and the nature of the sites. In every reaction catalyzed by acids, Brønsted acidity plays an important role. Nonetheless, the mechanistic information on surfaces of carbenium and carbonium ion are different due to of the deficiency of stabilizing solute species in the solvent (solvation effect) in heterogeneously reactions. The prime differences in their acid strength are believed to be a cause by the energetic positions of their electronically excited heterolytic terms which caused a formation of a stable covalent alkoxides. Locale of OH groups basic sites could take place in the process, thus are not promising to be reaction intermediates in system, but rather excited unstable ion pairs or transition states effect from excitation state of covalent surface alkoxy species. Due to this double functional nature of active acid sites in heterogeneous acid catalysis, it is essential to determine the solid catalysts acidic and basic properties.

1.10   Deactivation of Catalyst

The deactivation of catalyst can cause degradation of activity as well as in selectivity. Even though the degradation activities can be enhanced by elevating the temperature reaction, it is not practical enough for the generation of catalyst. Types of catalyst deactivation occur from poisoning, fouling, thermal degradation and active component volatility (Bartholomew 2001).

Poisoning

Impeding of active sites by specific elements or compounds supported by chemisorption are the chief causes of catalyst deactivation. If the formation of surface complexes is weak, reactivation may prevail, however if a formation is strong, results in deactivation. Poisonous chemical species are found in five classes which are Group 15 and 16 elements, metals and ions, molecules with free electron pairs that are strongly chemisorbed, and other compounds which can react with different active sites (Bartholomew 2001).

Fouling

Fouling occurs when the majority of catalysts and supports are porous. This blockage of pores by polymeric compounds typically affect the deactivation of the catalyst. At increasing temperatures (>770 K), the materials are transformed to black carbonaceous materials generally called coke (Thevenin et al. 2001). Catalysts having acidic or hydrogenating—dehydrogenating processes are especially sensitive to coking.

Thermal Degradation

An agglomeration of small metal crystallites below the melting point is an example of thermal degradation (Argyle and Bartholomew 2015). When the rate of sintering increases in elevated temperature, the presence of steam in the feed can rapid the sintering of metal crystallites thus causing deactivation of the catalyst. Besides, solid-solid reactions involving higher temperatures is another type of thermal degradation. Reactions between metals, such as Cu and alumina support resulted in the production of inactive metal aluminates (Bartholomew 2001) where degradation of activity and mechanical strain in catalyst pellets are severe.

Active Components Volatality

Degradation of activity of some catalytic systems containing P_2O_5, MoO_3, and many more during heating processes close to the sublimation point. Noble metals and Cu, Ni, Fe can easily be evaporated from catalysts after conversion to volatile chlorides and if traces of chlorine are present in the system.

1.11   Regeneration of Catalyst

The regeneration of metallic catalysts poisoned by Group 16 elements is rather laborious. Oxidation of sulfur by catalyst forms the corresponding sulfates (SO_3). However, if the catalyst is brought on-line under reducing conditions, then H_2S is formed poisoning the metallic catalysts (Bartholomew 2001). Thus, removal of poisons from the system is required and further prevention can be done by installing a guard-bed comprising of effective poison adsorbents in front of the reactor. Nickel catalysts poisoned with carbon monoxide can be regenerated by hydrogen treatment at suitable temperatures. Acidic catalysts poisoned partially by water, alcohols, or ammonia can be reinstated by thermal treatment at ample temperatures. In the point of view of industrial applications, the regeneration of coked catalysts plays a crucial role. The removal of coke can be done in a controlled temperature with oxygen, steam, and nitrogen atmospheric conditions in the presence of alkali metals (e.g., K) to speed up the coke gasification. Regeneration of catalysts that are deactivated by thermal degradation are very demanding. Some $Pt-Al_2O_3$ catalysts deactivated as a result during sintering. Regeneration can be done by chlorine treatment at elevated temperatures, which makes platinum redistribution possible.

References

Alborn TL (1989) Negotiating notation: chemical symbols and british society, 1831–1835. Ann Sci 46(5):437–460

Argyle MD, Bartholomew CH (2015) Heterogeneous catalyst deactivation and regeneration: a review. Catalysts 5(1):145–269

Auroux A, Gervasini A (2003) Infrared spectroscopic study of the acidic character of modified alumina surfaces. Adsorpt Sci Technol 21(8):721–737

Bartholomew CH (2001) Mechanisms of catalyst deactivation. Appl Catal A 212(1):17–60

Bartholomew CH, Farrauto RJ (2011) Fundamentals of industrial catalytic processes. Wiley, USA

Borm P, Klaessig FC, Landry TD, Moudgil B, Pauluhn J, Thomas K, Trottier R, Wood S (2006) Research strategies for safety evaluation of nanomaterials, part V: role of dissolution in biological fate and effects of nanoscale particles. Toxicol Sci 90(1):23–32

Borodziński A, Cybulski A (2000) The kinetic model of hydrogenation of acetylene–ethylene mixtures over palladium surface covered by carbonaceous deposits. Appl Catal A 198(1): 51–66

Boudart M (1985) Heterogeneous catalysis by metals. J Mol Catal 30(1–2):27–38

Burke MS, Kast MG, Trotochaud L, Smith AM, Boettcher SW (2015) Cobalt–iron (oxy) hydroxide oxygen evolution electrocatalysts: the role of structure and composition on activity, stability, and mechanism. J Am Chem Soc 137(10):3638–3648

Busca G (2007) Acid catalysts in industrial hydrocarbon chemistry. Chem Rev 107(11):5366–5410

Campanati M, Fornasari G, Vaccari A (2003) Fundamentals in the preparation of heterogeneous catalysts. Catal Today 77(4):299–314

Che M (2013) Nobel Prize in chemistry 1912 to sabatier: organic chemistry or catalysis? Catal Today 218:162–171

Chiesa M, Giamello E, Che M (2010) EPR characterization and reactivity of surface-localized inorganic radicals and radical ions. Chem Rev 110(3):1320

Clark JH (2002) Solid acids for green chemistry. Acc Chem Res 35(9):791–797

Clarke JK, Creaner AC (1981) Advances in catalysis by alloys. Industrial & Engineering Chemistry Product Research and Development 20(4):574–593

Clugston M, Flemming R (2000) Advanced chemistry. Oxford University Press, Oxford

Coq B, Figueras F (2001) Bimetallic palladium catalysts: influence of the co-metal on the catalyst performance. J Mol Catal A Chem 173(1):117–134

Cotton FA, Wilkinson G, Murillo CA, Bochmann M, Grimes R (1988) Advanced inorganic chemistry, vol 5. Wiley, New York

De Jong KP, Geus JW (2000) Carbon nanofibers: catalytic synthesis and applications. Catal Rev 42(4):481–510

De Vos D, Thibault-Starzyk F, Knops-Gerrits P, Parton R, Jacobs P (1994) A critical overview of the catalytic potential of zeolite supported metal complexes. In: Macromolecular symposia, vol 1. Wiley Online Library, pp 157–184

De Vos DE, Dams M, Sels BF, Jacobs PA (2002) Ordered mesoporous and microporous molecular sieves functionalized with transition metal complexes as catalysts for selective organic transformations. Chem Rev 102(10):3615–3640

Dubois LH, Nuzzo RG (1992) Synthesis, structure, and properties of model organic surfaces. Annu Rev Phys Chem 43(1):437–463

Eckert H, Wachs IE (1989) Solid-state vanadium-51 NMR structural studies on supported vanadium (V) oxide catalysts: vanadium oxide surface layers on alumina and titania supports. J Phys Chem 93(18):6796–6805

Ertl G, Knözinger H, Weitkamp J (2008) Preparation of solid catalysts. Wiley, USA

Fresco LO (2015) The new green revolution: bridging the gap between science and society. Curr Sci 109(3):430–438

Gai PL (1999) Environmental high resolution electron microscopy of gas-catalyst reactions. Top Catal 8(1):97–113

Geiger FM (2009) Second harmonic generation, sum frequency generation, and χ(3): dissecting environmental interfaces with a nonlinear optical Swiss Army knife. Annu Rev Phys Chem 60:61–83

Goodman DW (1996) Correlations between surface science models and "real-world" catalysts. J Phys Chem 100(31):13090–13102

Govindasamy A, Muthukumar K, Yu J, Xu Y, Guliants VV (2010) Adsorption of propane, isopropyl, and hydrogen on cluster models of the M1 phase of Mo–V–Te–Nb–O mixed metal oxide catalyst. J Phys Chem C 114(10):4544–4549

Grasselli F, Basini G, Bussolati S, Bianco F (2005) Cobalt chloride, a hypoxia-mimicking agent, modulates redox status and functional parameters of cultured swine granulosa cells. Reprod Fertil Dev 17(7):715–720

He Q (2013) Study of heterogeneous gold and gold alloy catalysts via analytical electron microscopy. Lehigh University, USA

Hutchings GJ (2001) Promotion in heterogeneous catalysis: a topic requiring a new approach? Catal Lett 75(1):1–12

Kear B, Breinan E, Greenwald L (1979) Laser glazing–a new process for production and control of rapidly chilled metallurgical microstructures. Met Technol 6(1):121–129

Kemball C, Dowden D (1981) Catalysis, vol 4. Royal Society of Chemistry, UK

Korlach J, Turner S (2007) Articles having localized molecules disposed thereon and methods of producing and using same. Google Patents

Kozuch S, Martin JM (2012) "Turning over" definitions in catalytic cycles. American Chemical Society, USA

Kreuer K (1997) On the development of proton conducting materials for technological applications. Solid State Ionics 97(1):1–15

Lazcano A, Peretó J (2010) Should the teaching of biological evolution include the origin of life? Evol Educ Outreach 3 (4):661–667

Lundie DT, McInroy AR, Marshall R, Winfield JM, Jones P, Dudman CC, Parker SF, Mitchell C, Lennon D (2005) Improved description of the surface acidity of η-alumina. J Phys Chem B 109 (23):11592–11601

Mallat T, Baiker A (2000) Selectivity enhancement in heterogeneous catalysis induced by reaction modifiers. Appl Catal A 200(1):3–22

Mayer KM, Hafner JH (2011) Localized surface plasmon resonance sensors. Chem Rev 111 (6):3828–3857

Moulijn JA, Van Diepen A, Kapteijn F (2001) Catalyst deactivation: is it predictable? What to do? Appl Catal A 212(1):3–16

Mukhopadhyay K, Mandale AB, Chaudhari RV (2003) Encapsulated HRh (CO)(PPh3) 3 in microporous and mesoporous supports: novel heterogeneous catalysts for hydroformylation. Chem Mater 15(9):1766–1777

Nakamura H, Matsui Y (1995) Silica gel nanotubes obtained by the sol-gel method. J Am Chem Soc 117(9):2651–2652

Nijhuis TA, Beers AE, Vergunst T, Hoek I, Kapteijn F, Moulijn JA (2001) Preparation of monolithic catalysts. Catal Rev 43(4):345–380

Olah GA, Molnar A (2003) Hydrocarbon chemistry. Wiley, USA

Paulik MG, Brooksby PA, Abell AD, Downard AJ (2007) Grafting aryl diazonium cations to polycrystalline gold: insights into film structure using gold oxide reduction, redox probe electrochemistry, and contact angle behavior. The Journal of Physical Chemistry C 111 (21):7808–7815

Pelaez M, Nolan NT, Pillai SC, Seery MK, Falaras P, Kontos AG, Dunlop PS, Hamilton JW, Byrne JA, O'shea K (2012) A review on the visible light active titanium dioxide photocatalysts for environmental applications. Appl Catal B 125:331–349

Peyratout CS, Daehne L (2004) Tailor-made polyelectrolyte microcapsules: from multilayers to smart containers. Angew Chem Int Ed 43(29):3762–3783

Price PM, Clark JH, Macquarrie DJ (2000) Modified silicas for clean technology. J Chem Soc Dalton Trans 2:101–110

Ramírez J, Macías G, Cedeño L, Gutiérrez-Alejandre A, Cuevas R, Castillo P (2004) The role of titania in supported Mo, CoMo, NiMo, and NiW hydrodesulfurization catalysts: analysis of past and new evidences. Catal Today 98(1):19–30

Raybaud P (2007) Understanding and predicting improved sulfide catalysts: Insights from first principles modeling. Appl Catal A 322:76–91

Rimola A, Costa D, Sodupe M, Lambert J-F, Ugliengo P (2013) Silica surface features and their role in the adsorption of biomolecules: computational modeling and experiments. Chem Rev 113(6):4216–4313

Roberts FS, Anderson SL, Reber AC, Khanna SN (2015) Initial and Final State Effects in the Ultraviolet and X-ray Photoelectron Spectroscopy (UPS and XPS) of Size-Selected Pd n Clusters Supported on TiO2 (110). J Phys Chem C 119(11):6033–6046

Ryczkowski J (2001) IR spectroscopy in catalysis. Catal Today 68(4):263–381

Sachtler W (1984) Selectivity and rate of activity decline of bimetallic catalysts. J Mol Catal 25(1–3):1–12

Senkan S (2001) Combinatorial heterogeneous catalysis—a new path in an old field. Angew Chem Int Ed 40(2):312–329

Skjærvø Ø (2016) Characterization of open tubular enzyme reactors and polymer layer open tubular columns for liquid chromatography. MS thesis

Smil V (2004) Enriching the earth: Fritz Haber, Carl Bosch, and the transformation of world food production. MIT press

Stone FS (1975) The significance for oxide catalysis of electronic properties and structure. J Solid State Chem 12(3–4):271–281

Stranges AN (1984) Friedrich Bergius and the rise of the German synthetic fuel industry. Isis 75 (4):643–667

Sturgeon MR, O'Brien MH, Ciesielski PN, Katahira R, Kruger JS, Chmely SC, Hamlin J, Lawrence K, Hunsinger GB, Foust TD (2014) Lignin depolymerisation by nickel supported layered-double hydroxide catalysts. Green Chem 16(2):824–835

Tanaka M, Itadani A, Kuroda Y, Iwamoto M (2012) Effect of pore size and nickel content of Ni-MCM-41 on catalytic activity for ethene dimerization and local structures of nickel ions. J Phys Chem C 116(9):5664–5672

Thevenin P, Ersson A, Kušar H, Menon P, Järås SG (2001) Deactivation of high temperature combustion catalysts. Appl Catal A 212(1):189–197

Thomas JM, Raja R, Lewis DW (2005) Single-Site heterogeneous catalysts. Angew Chem Int Ed 44(40):6456–6482

Tossell JA (1975) The electronic structures of Mg, Al and Si in octahedral coordination with oxygen from SCF Xα MO calculations. J Phys Chem Solids 36(11):1273–1280

Valigi M, Gazzoli D, Cimino A, Proverbio E (1999) Ionic size and metal uptake of chromium (VI), molybdenum (VI), and tungsten (VI) species on ZrO2-based catalyst precursors. J Phys Chem B 103(51):11318–11326

van Santen RA (2017) Modern heterogeneous catalysis: an introduction. Wiley, USA

Walawalkar MG, Roesky HW, Murugavel R (1999) Molecular phosphonate cages: model compounds and starting materials for phosphate materials. Acc Chem Res 32(2):117–126

West AR (2007) Solid state chemistry and its applications. Wiley, USA

Xue C-H, Ma J-Z (2013) Long-lived superhydrophobic surfaces. J Mater Chem A 1(13): 4146–4161

Xuereb DJ, Raja R (2011) Design strategies for engineering selectivity in bio-inspired heterogeneous catalysts. Catal Sci Technol 1(4):517–534

Zhang Z, Wong TT, Sachtler WM (1991) The effect of Ca^{2+} and Mg^{2+} ions on the formation of electron-deficient palladium-proton adducts in zeolite Y. J Catal 128(1):13–22

Zhidomirov GM, Chuvylkin ND (1986) Quantum-chemical methods in catalysis. Russ Chem Rev 55(3):153–164

Zhu Z, Radovic L, Lu G (2000) Effects of acid treatments of carbon on N_2O and NO reduction by carbon-supported copper catalysts. Carbon 38(3):451–464

Chapter 2
Design of Fuel Cells and Reactors, Estimation of Process Parameters, Their Modelling and Optimization

2.1 Introduction

Fuel cell systems are a promising alternative power source based technologies that are vastly used in today's industrial applications. Fuel cells are clean, quiet, and feasible devices that transformed chemical changes into electricity to be utilized. They are operational and can generate fuel as long as it is supplied with oxygen and hydrocarbons. Evidently, emission such as carbon dioxide are greatly reduced mainly by fuel cell technologies compared to conventional technologies (Höök and Aleklett 2010). Besides, fuel cells are ideal for power generation to provide supplemental power and backup assurance for critical areas, or for on-site service in areas that are inaccessible by power lines. Since fuel cells operate silently, they reduce noise pollution as well as air pollution and the waste heat from a fuel cell can be used to provide hot water or space heating. Fuel cells are the best alternative because they combine higher fuel efficiency with low or no pollution, greater flexibility in installation and operation, quiet operation, low vibration, and potentially lower maintenance and capital costs (Stambouli and Traversa 2002). The wide range of applications includes unmanned under water vehicle, automotive, locomotives, surface ships, electronic component and automobiles (Woodland 2001). Fuel cells are a propitious alternative energy conversion technology. Typical fuel cell utilizes hydrogen as a fuel and reacting it with oxygen to produce electricity. The fundamental fabrication of these fuel cells is its ability to employ the equations of standard reactor design to the reaction kinetics of a fuel cell as shown below (Figs. 2.1 and 2.2).

$$Anode: \text{H}_2 + \text{O}^{2-} \rightarrow \text{H}_2\text{O} + 2\text{e}^-$$

$$Cathode: \frac{1}{2}\text{O}_2 + 2\text{e}^- \rightarrow \text{O}^{2-}$$

$$Overall: \text{H}_2 + \frac{1}{2}\text{O}_2 \rightarrow \text{H}_2\text{O}$$

© Springer International Publishing AG 2017
S. Bagheri, *Catalysis for Green Energy and Technology*,
Green Energy and Technology, DOI 10.1007/978-3-319-43104-8_2

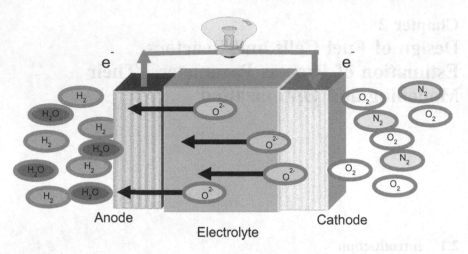

Fig. 2.1 Reactions of a typical fuel cell

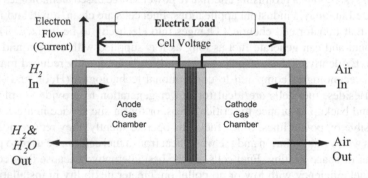

Fig. 2.2 Flow diagram for a fuel cell system

In this reaction, when each mole of hydrogen used up, two moles of electrons moved through the electric load. Conversion of electron flow can be done by applying Faraday's constant ($F = 96,485$ coulombs/mole of electrons) and mathematical calculation of energy utilizes. The potential of a fuel cell usually relates in terms of efficiency, of the energy available from the reaction in the fuel cell system. A recent study in polymer electrolyte membrane (PEM) and catalyst technology have increased the fuel cell power density and made them available for automobiles and portable power applications, as well in power plants. The negative potential side of the cell stack is fed by the fuel processing system (FPS) which convert natural gas or other hydrocarbons into clean, purity and rich hydrogen and by-products. On the other hand, the positive potential is supplied with oxygen gas. Both gases in anode and cathode reacts electrochemically with each other to generate electricity.

2.2   Drawback in Fuel Cell Power Systems

The most complication in fuel cell power systems lies within the FPS to the anode site of the cell stack. Poor supply of oxygen causes "starvation" of the cell, which denotes that the platinum catalyst will start consuming the graphite used in the flow fields, thus resulting in fast and permanent harm (Ahn 2011; Mandal 2016; Chen and Hsu 2011). While insufficient supply of oxygen reduces the cell span life, too much hydrogen output from FPS lowers its efficiency, which is not promising. This is due to a high fuel flow in a substantial surplus of hydrogen, leaving the fuel cell results in a relatively high reformer temperature with high conversion efficiency, but with lower overall plant efficiency. This causes are resulted from excess amounts of hydrogen being burned rather than utilized in the conversion to electricity (Scheffler 1991). Another problem with fuel cells, especially PEM is to maintain the carbon monoxide level below a critical limit 10 ppm for PEM fuel cells (Ormerod 2003; Schmittinger and Vahidi 2008; Pettersson and Westerholm 2001). Carbon monoxide is harmful and easily produced during the reforming reaction, resulting in de-activation of the platinum electro catalyst. In the worst scenario, occurrence of carbon monoxide limits hydrogen from reacting in the anode side influencing the system efficiency and lowers the fuel cell voltage measurement.

Another crucial limitation is in the temperature of CPO, where it must be at a certain point (Linde 1983). Too high temperature damages CPO catalyst bed permanently while the temperature decreases lower down the fuel reaction rate (Landsman 1989). Not only that, the implementation of these designs is the absence of reliable measurements of hydrogen partial pressure in the cell stack. Existing sensors are not apt for use with these controllers because, as discussed, commercially available sensors are not fabricated to operate in a reformate gas environment. This is because the sensors are either too bulky or costly. Sensors that are affordable are not reliable or do not have long life time (Chong and Kumar 2003).

2.3   Design of Fuel Cells and Reactors

In a fuel system, generating chemical energy into electrical energy can be done by transforming hydrogen and oxygen into steam were restricted by efficiency limits of the Carnot thermal cycle (Larminie and Dicks 2000). Since then, many remarkable approaches have been investigated on fuel cell technology; however, it is not widely used in the automotive industry. This is because the impotent efficiency of the hydrogen distribution centre in the fuel cell system (Lovins and Williams 1999) as well as the trouble related to storing hydrogen on board an automobile. Thus, designing a fuel cell that employs carbon-based hydrogenous fuel are utilized as it widely available, aids in generating an in situ hydrogen and enables easy store on-board a vehicle (Dicks 1996).

The main components of a fuel cell power system comprise of two units that are the fuel cell stack and the auxiliary subsystem. In this context, the fuel cell stack is discussed as the core design which comprises of a reformer unit, a water shift gas reactor and a catalytic preferential oxidation reactor, which converts methane gas into hydrogen of the desired purity as shown in Fig. 2.3. This concept employs in a model of a polymer electrolyte membrane (PEM) fuel cell system and the quantity of power induced is reckoned as a function of hydrogen penetrating the fuel cell. Subsequently, the hydrogen penetrates into the fuel cell and intermingle with oxygen to promote electrical power which pilot an electric motor. However, in designing an ideal fuel cell, numerous operational constraints are to be taken note of such as the steam or autothermal reforming, water gas shift (WGS) reactor and catalytic preferential oxidation (PrOx) reactor.

The reforming process function as cleaning of sulfur from natural hydrocarbon to avoid fuel cell anode electrode bane. Then, the steam reforming of methane enters the WGS reactor, which was designed as a packed bed plug flow reactor to inhibit all exothermic reactors. When the temperature is high, the preliminary reaction rate is rapid but the equilibrium conversion is low. On the other hand, when the temperature is low, the reaction rate decreases; however, the equilibrium conversion is high. Therefore, water gas shift reactor is typically split into two zones with a high temperature zone (where the rate of reaction is low) and a lower temperature zone (where the rate of reaction is high). Following, is the catalytic preferential oxidation (PrOx) reactor, which was designed as an isothermal packed bed reactor to avoid poisoning of the PEM system from the output of water gas shift reactor due to carbon monoxide gas (Godat and Marechal 2003).

Lastly, the effect of manipulating the feed rate of methane on the overall hydrogen generated with the same reactor train was scrutinized. The methane supplied to the reformer differ between 0.0163 and 0.167 mol/s. The steam flow rate was calibrated in order to maintain the steam to methane ratio at 3:1 for all studies. The correlation between hydrogen egress from the catalytic partial oxidation reactor and the methane infiltrate into the reformer is as shown in Eq. 2.1.

$$N \cdot_{H_2} = 3.12 N \cdot_{CH_4} \qquad (2.1)$$

From Eq. 2.1, a fitting straight line can be plotted in a steady state of hydrogen production rate against the methane production is represented in Fig. 2.4. This result displayed that the ideal fuel cell can be operated near to equilibrium.

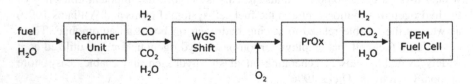

Fig. 2.3 A simple designated fuel cell power system (Zalc et al. 2002)

Fig. 2.4 A fitting straight line can be plotted in steady state of hydrogen production rate against the methane production (Bordo and Murshid 2006)

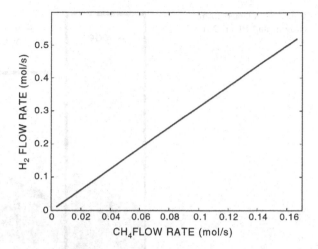

2.4   Estimation of Process Parameters in Fuel Cells

One of the main limitation of the fuel cell is its cost, efficiency, size and maintaining the auxiliary part in order to sustain the physical properties of fuels. The modelling of a prototype fuel cell is typically associated with the reactor operation to the system parameters, such as, concentrations of the reactant, loading of the catalyst, pressure, temperature, and etc. On the other hand, temperature, pressure, flow rate, composition, and geometry are coordinated with heat and mass transfer. The mathematical association between the parameters and unknown are the prime constituent in providing the estimation optimum performance information in engineering the fuel cell system.

One of the parameters to be taken note are the water dynamics in the fuel cell stack (Kazmi and Bhatti 2012). Water dynamics play a role as it comprises of electro-osmosis flow from the negatively charged electrode to the positively charged electrode and diffuse back to and fro across the membrane as shown in Fig. 2.5.

This is because the existence of water contents of the stack is vulnerable to length of membrane life, cold environment, proton conductivity and reactant gas transportation. For instance, in a cold environment, the water content in the stack can interchange phase due to subfreezing temperature. Besides, lacking or too much of water in the stack can greatly decline the performance of the system. When the amount of water is less, it can cause the membrane to dry up and this would form cracks in the membrane which eventually resulted in a reduction in the performance and short life span of the system. In the latter part, too much water in the stack prompt to flooding the pores of the porous gas diffusion layer and advert the dispersal of the reactants to the catalyst. Subsequently, the produced water at the positive electrode diffuses into the negative electrode due to the gradient of water concentration. This resulted in severe problems where water will be accumulated

Fig. 2.5 Adapted from
(Kazmi and Bhatti 2012)

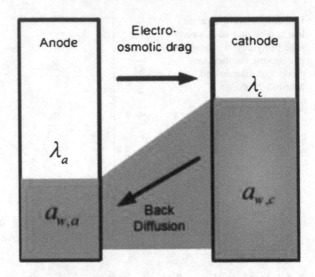

and obstruct the sufficient dispersal of fuel gas. In addition, other dynamical behaviour such as listed below (A-G) are evaluated and adequately modelled to ensure the quality of fuel needed by fuel cells.

(A) Cell Reversible Voltage (E_{oc})

The ideal reversible/open circuit voltage for a fuel cell is defined as electrical work done in moving per unit charge on one mole of electrons through the fuel cell circuit in Eq. 2.2.

$$E_{OC} = W_{el}/2F \qquad (2.2)$$

where, F is the Faraday constant, W_{el} is the electrical work and "2" in the denominator denotes the number of electrons that flow for one mole of hydrogen. In an ideal system, it is derived from an alteration of Nernst equations assessing from the feasible variations of temperature from the standard value of 25 °C.

$$E_{Nerst} = 1.229 - 0.85^{10^{-3}}(T - 298.15) + 4.31^{10^{-5}}T[\ln(P_{H_2}) + 1/2\ln(P_{O_2}) \quad (2.3)$$

whereby T is the temperature of cell operation in [K], P_{H2} and P_{O2} are respectively the partial pressures of hydrogen and oxygen in [atm].

(B) Coefficient of Fuel Utilization (μF)

In experimental, hydrogen that enters the fuel cell is not entirely used in the electrochemical reaction. Therefore, the fuel utilization coefficient, μF, is as shown in Eq. 2.4 below:

$$\mu F = \text{Mass of fuel reacted in cell/Mass of fuel input to cell} \qquad (2.4)$$

The mass of fuel responded in the fuel cell is ameliorate with high fuel percentage of hydrogen (Ralph et al. 1997).

(C) Efficiency of Fuel Cell (η):

The fuel cell efficiency is influenced by the actual voltage produced in the fuel cell. V_{out}, the actual fuel cell output voltage, are as in Eq. 2.5.

$$V_{out} = E_{oc} - V_{drop} \qquad (2.5)$$

where, V_{drop} is, the voltage drop within the fuel cell. The cell efficiency η is then written as in Eq. 2.6:

$$\eta = (\mu F \ VC)/E_{oc} \qquad (2.6)$$

whereby the voltage drop in the system is predominantly due to polarization losses, which involves activation polarization, concentration polarization and losses of ohmic polarization (Gür et al. 1980).

(D) Consumption of Hydrogen

The hydrogen consumption in a fuel cell is determined by the type of fuel cell and the concentration of hydrogen at standard temperature and pressure as in Eq. 2.7 below.

$$H_2 \ consumption = (2.02 \times 10^{-3} P_{out})/(2V_c F) \qquad (2.7)$$

whereby 2.02×10^{-3} kg/mole is the molar mass of hydrogen at STP and P_{out} is the electrical power output of the fuel cell.

(E) Heating Rate of Fuel Cell

The ideal E_{oc} is produced only if the whole heat energy of combustion in the fuel cell is transformed into electrical energy. However, in practical, some heat energy may have lost in the by products that caused from the electrochemical reactions at the positive and negative electrode. Thus, by using the net heat produced during combustion (LHV) value, the E_{oc} for a fuel cell is 1.25 V (Gomaton and Jewell 2003).

$$Heating \ rate = nI(1.25 - V_c)W \qquad (2.8)$$

whereby I, is the maximum current for a stack of n cells.

(F) Net Power Output (NP_{out})

The net power output (NP_{out}) of fuel cell is the available electric power output to the connected load. The net power output is equivalent to the electrical power output

P_{out} deducted the total parasitic power and conversion losses. The subsystems in a fuel cell unit rely on the operating temperature range, the type of the fuel cell, and the nature of electrochemical reactions at both terminal electrode with the Eq. 2.9 as below.

$$NP_{out} = P_{out} - \sum(\text{parasitic loss} + \text{conversion loss}) \qquad (2.9)$$

(G) Total Efficiency (η tot)

The total efficiency of the fuel cell generator system, η tot, is the ratio of the total net power output and the net heat released in the exhaust to the total system of net heat production during combustion of the fuel input as written Eq. 2.10:

$$\eta\, tot = (NP_{out} + H_{exh})/T_{LHV} \qquad (2.10)$$

whereby H_{exh} is the net heat released in the exhaust and T_{LHV} is the total system LHV fuel input.

2.5 Modelling and Optimization for Fuel Cells

There are two distinct methods in modelling and optimizing the fuel stack cell; by static or dynamic behaviour. The latter modelling, which applies Spiros stack model shows that understanding the dynamic behaviour is the key to the fuel cell system in the time domain to the changing of the load. Spiros' model utilizes large mass of loads with uniform environment inside the reactors making the assumption sensible due to the exit values of mole fractions of interest. However, accuracy of mass models can be greatly influenced by thermal waves and reaction zones which may cause variability in modelling model and inscribed with rigidity of the tools. Thus, the parameters such as the catalytic partial oxidation, water gas shift and catalytic preferential oxidation in the irreversibility of the fuel stack cell needed extended measurements.

2.5.1 Catalytic Partial Oxidation (CPO)

In the CPO, there are two main reactions that plays major role:

(i)   *Partial Oxidation (Pox)*: $CH_4 + 1/2O_2 \rightarrow CO + 2H_2$ \qquad (2.11)

(ii)   *Full Oxidation (Fox)*: $CH_4 + 2O_2 \rightarrow CO_2 + 2H_2O$ \qquad (2.12)

In partial oxidation, it generates essential hydrogen gases, but also carbon monoxide, which prompts to poisoning in the cell stack. On the other hand, full oxidation occurs by supplying additional heat to the system which generates the feasibility of a partial oxidation reaction. The rate of reaction for full-and partial-oxidation are depicted in the Equation below:

$$Rate\ of\ Pox = vrt \tag{2.13}$$

$$Rate\ of\ Fox = (1 - v)rt \tag{2.14}$$

where, v is a selectivity variable (Dicks 1996; Ormerod 2003) which are influenced by a ration of air-fuel and rt is the summation of reaction rate shown in Eq. 2.15:

$$rt = kg[O_2] \frac{[CH_4]}{[CH_4] + \varepsilon} \tag{2.15}$$

whereby $[O_2]$ and $[CH_4]$ indicates the concentrations of oxygen and methane and kg and ε are coefficients from empirical studies. The term, $kg[O_2]$, in Eq. 2.15 denotes the mass transfer rate of oxygen from gas phase to the catalyst. The second term, $[CH_4]/[CH_4] + \varepsilon$ is a function that reckons for the position in which methane is the limiting reactant.

The dynamic model is derived from mole balance equations as shown below

$$M = u_f + u_a - F_{out}^{CPO} + l_1 r_{p_{ox}} V + l_2 r_{F_{ox}} V \tag{2.16}$$

where, u_f, u_a and F_{out}^{CPO} (mole/sec) are the fuel, air and exit molar flow vectors, V is the reactor volume in meter per cubic, and l_1 and l_2 are stoichiometric coefficient vectors where;

$$l_1 = [0 \quad -1 \quad 1 \quad 0 \quad 2 \quad -1/2] \tag{2.17}$$

$$l_2 = [0 \quad -1 \quad 0 \quad 1 \quad 0 \quad 2 \quad -2] \tag{2.18}$$

Thus, the energy balance principle of dynamics given temperature T is as shown in Eq. 2.19.

$$\left(m_{c_p}\right)T - u_f h\left(T_f\right) + u_a h(T_a) - F_{out}^{CPO} hT + \delta H_{Pox} r_{Pox} V + \delta H_{Fox} r_{Fox} V \tag{2.19}$$

whereby T is the reaction temperature in Kelvin, m is the mass in kg, cp (kJ/kg K) a is the specific heat capacity of the catalyst bed. The terms $h(T_f)$, $h(T_a)$ and $h(T)$ are the molar enthalpies of ideal gas for fuel, air and the exit temperatures respectively and δ H is the reaction heat at reference temperature.

2.5.2   Water Gas Shift (WGS)

This system comprises of two inlet streams in which (1) hydrocarbon reforming from CPO and (2) feed from the reservoir. In the reactor carbon monoxide reacts with water and generates hydrogen and carbon dioxide. Thus, an expression on rate of reaction for WGS, from the principle of Mass Action Law and the Arrhenius Equation is expressed in Eq. 2.20.

$$r = M_f e^{\frac{K_f}{RT}}[CO][H_2O] - M_b e^{\frac{K_b}{RT}}[CO_2][H_2] \tag{2.20}$$

where, M_f, M_b, K_f and K_b are parameters of the reaction rate.

Conveying it in terms of expressing the mole balance equations, M, below is as in Eq. 2.21:

$$M = u_g + u_w - F_{out}^{WGS} + krV \tag{2.21}$$

whereby u_g and u_w, are the flow vectors of gas, air and F_{out}^{WGS} is the exit molar flow vectors and q is a constant with [0 0 −1 1 1 −1 0].

Subsequently, the dynamics of the temperature T are equated as below:

$$(m_{cp})T = u_g h(T_g) + u_w h(T_w) - F_{out}^{WGS} h(T) + \delta H_r V \tag{2.22}$$

Inscribe that the dynamic variables, M and T and constants, V, m, cp and δ H have been used in the reactor system CPO, WGS and PrOx as expressed.

2.5.3   Catalytic Preferential Oxidation (PrOx)

The PrOx is the last stage in eliminating poisonous carbon monoxide such as proton exchange membrane in fuel cell into carbon dioxide and water as expressed in Eqs. 2.23 and 2.24

$$CO + 1/2\,O_2 \rightarrow CO_2 \tag{2.23}$$

$$H_2 + 1/2\,O_2 \rightarrow H_2O \tag{2.24}$$

Apt expressions of the rate of reaction are expressed in equations below,

$$r_{CO} = s^{PrOx} r_t^{PrOx} \tag{2.25}$$

$$r_{H_2} = \left(1 - s^{PrOx}\right) r_t^{Prox} \tag{2.26}$$

where, s^{PrOx} s is a selectivity variable. Therefore, the total rate of reaction can be equated as:

$$r_t^{PrOx} = k[CO][O_2]^{1/2} \tag{2.27}$$

Then,

$$M = u_g^{PrOx} + u_a^{PrOx} - F_{out}^{PrOx} + m_1^{PrOx} r_{CO} V + m_2^{PrOx} r_{H_2} V \tag{2.28}$$

where, u_g^{PrOx}, u_a^{PrOx},. and F_{out}^{PrOx} are the molar flow vectors for gas, air and exit respectively, and given, $m_1^{PrOx} = [0\ 0\ -1\ 1\ 0\ 0\ -1/2]$ and $m_2^{ProX} = [0\ 0\ 0\ 0\ -1\ 1\ -1/2]$.

Similarly, temperature dynamics are given by

$$(mc_p) T = u_g^{PrOx} k(T_g) + u_a^{PrOx} k(T_a) - F_{out}^{PrOx} h(T) + \delta H_{CO} r_{CO} V + \delta H_{H_2} r_{H_2} V \tag{2.29}$$

References

Ahn J-W (2011) Control and analysis of air, water, and thermal systems for a polymer electrolyte membrane fuel cell. Auburn University, Alabama

Bordo MD, Murshid AP (2006) Globalization and changing patterns in the international transmission of shocks in financial markets. J Int Money Finance 25(4):655–674

Chen Z, Hsu R (2011) Catalyst support degradation. In: PEM fuel cell failure mode analysis. CRC Press, Boca Raton, pp 33–72

Chong C-Y, Kumar SP (2003) Sensor networks: evolution, opportunities, and challenges. Proc IEEE 91(8):1247–1256

Dicks AL (1996) Hydrogen generation from natural gas for the fuel cell systems of tomorrow. J Pow Sources 61(1–2):113–124

Godat J, Marechal F (2003) Optimization of a fuel cell system using process integration techniques. J Pow Sources 118(1):411–423

Gomaton P, Jewell W (2003) Fuel parameter and quality constraints for fuel cell distributed generators. In: 2003 IEEE PES transmission and distribution conference and exposition. IEEE, pp 409–412

Gür TM, Raistrick ID, Huggins RA (1980) Steady-State D-C Polarization Characteristics of the O₂, Pt/Stabilized Zirconia Interface. J Electrochem Soc 127(12):2620–2628

Höök M, Aleklett K (2010) A review on coal-to-liquid fuels and its coal consumption. Int J Energy Res 34(10):848–864

Kazmi IH, Bhatti AI (2012) Parameter estimation of proton exchange membrane fuel cell system using sliding mode observer. Int J Innovative Comput Inform Control 8(7B):5137–5148

Landsman N (1989) Limitations to dimensional reduction at high temperature. Nucl Phys B 322 (2):498–530

Larminie J, Dicks A (2000) Fuel Systems Explained. Wiley, Chichester

Linde AD (1983) Decay of the false vacuum at finite temperature. Nucl Phys B 216(2):421–445

Lovins AB, Williams BD (1999) A strategy for the hydrogen transition. Rocky Mountain Institute Snowmass, Colorado

Mandal P (2016) Investigation and mitigation of degradation in hydrogen fuel cells

Ormerod RM (2003) Solid oxide fuel cells. Chem Soc Rev 32(1):17–28

Pettersson LJ, Westerholm R (2001) State of the art of multi-fuel reformers for fuel cell vehicles: problem identification and research needs. Int J Hydrogen Energy 26(3):243–264

Ralph T, Hards G, Keating J, Campbell S, Wilkinson D, Davis M, St-Pierre J, Johnson M (1997) Low cost electrodes for proton exchange membrane fuel cells performance in single cells and Ballard stacks. J Electrochem Soc 144(11):3845–3857

Scheffler GW (1991) Fuel cell power plant fuel control. Google Patents

Schmittinger W, Vahidi A (2008) A review of the main parameters influencing long-term performance and durability of PEM fuel cells. J Pow Sources 180(1):1–14

Stambouli AB, Traversa E (2002) Solid oxide fuel cells (SOFCs): a review of an environmentally clean and efficient source of energy. Renew Sustainable Energy Rev 6(5):433–455

Woodland RLK (2001) Autonomous marine vehicle. Google Patents

Zalc J, Sokolovskii V, Löffler D (2002) Are noble metal-based water–gas shift catalysts practical for automotive fuel processing? J Catalysis 206(1):169–171

Chapter 3
Catalysis in Fuel Cells (PEMC, SOFC)

3.1 Introduction

Energy is the basis of economic development, there is no modern civilization without the development of the energy industry. Humans have been conducting efforts to improve the high efficiency use of energy resources. There has been a number of revolutionary changes in the way to use energy during the history, from the original steam engine to internal combustion engines. Fuel cells are energy devices which transfer chemical energy stored in the fuel and oxidant directly into electrical energy. When fuel cells are continuously supplied fuel and oxidant, electricity can be made constantly. According to the different electrolytes, fuel cells can be divided into several types, such as alkaline fuel cell (AFC), phosphoric acid fuel cell (PAFC), molten carbonate fuel cell (MCFC), solid oxide fuel cell (SOFC), and proton exchange membrane fuel cell (PEMFC), etc. Fuel cells can supply electrical power in centralized or individual ways have advantages of high efficiency of energy conversion, clean, no pollution, low noise, modular structure (Pauniaho et al. 2014). Fuel cells are a significant technology for households and commercial buildings, plus, they are supplemental or auxiliary power to support cars, trucks and aircraft systems. Besides, fuel cells are the power for personal and commercial transportation of new power generation closely tailored to meet growth in power consumption. These applications will be used in a large number of industries worldwide. A large power plant such as thermal power plant can obtain high efficiency. However, giant units of the power plant are affected by various limitations, so customers can achieve power from an electrical grid in which the power plant makes concentrated power. The big unit's inflexible generation cannot adapt to the family's needs; besides, the grid sometimes goes to the peak or downturn with the user's electrical load change. In order to adapt to the change of electrical load, a water pumped-storage plant has been built for emergency, but it will reduce the benefit of the grid. The traditional coal-fired power plant's burning energy consumes nearly 70% in boilers and steam turbine generators, and a great

© Springer International Publishing AG 2017
S. Bagheri, *Catalysis for Green Energy and Technology*,
Green Energy and Technology, DOI 10.1007/978-3-319-43104-8_3

deal of hazardous substances is emitted out. However, fuel cells transfer chemical energy into electrical energy, without burning and rotating components.

Theoretically energy conversion closes to 100%. The actual power generation efficiency can reach 40–60%, energy can be transferred directly into enterprises, hotels, families. The comprehensive energy efficiency can reach 80%, and the devices are very flexible (An et al. 2007). Fuel cells are called the fourth electricity power generation after water, nuclear power generation devices. People from the international energy area forecast that the fuel cell is the most attractive power generation in the 21st century. It will bring huge economic benefits while combining traditional large units and power grid (Zhang et al. 2005).

3.2   Theory of Fuel Cell

In fuel cell, chemical energy stored in gaseous molecules of fuel is transformed into efficient electrical energy. This system can be equated by simple electrochemical reaction below:

$$2H_2 + O_2 \rightarrow 2H_2O \tag{3.1}$$

During combustion, hydrogen molecules are being oxidized whereas oxygen is being reduced releases free electrons. The 'free' electron reconfigure into the essential supply of electricity through an elementary circuit with a load in a consumable form. Simple fuel cells have a small contact between the electrolyte, the electrode and the gas fuel. The fuel cell unlike conventional battery does not require recharging as it will generate electricity as far as fuel is supplied. The fundamental modelling of fuel cell involves an electrolyte and two electrodes. Ions of hydrogen and oxygen transported over each of the electrodes respective through an electrolyte by means of a chemical reaction, by-products of water are produced along with generated electricity.

In Fig. 3.1, hydrogen and oxygen react to form chemical energy into efficient and cleaner electricity and by-products (heat and water). Hydrogen fuel is commonly found and utilizes in hydrocarbons, gasoline and carbon monoxide are as are simply hydrogen carriers. Hydrogen does not appear naturally in its pure form; however, it is formed through the steam reforming of natural gas. Nonetheless, it is costly as steaming-reforming of hydrogen are quite expensive. Another alternative is to generate hydrogen is electrolysis of solar hydrogen.

In Fig. 3.2, solar electric modules are employ to transfigure solar energy to the electric power, which can function as an electrolyzer. In the electrolyzer system, water molecules are electrolyzed to form hydrogen and oxygen gas. The gases will be liberated to the separated purifier and pure hydrogen will be kept in the hydrogen tank.

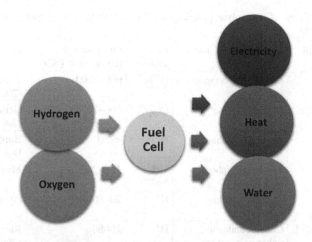

Fig. 3.1 The diagram of fuel cell inputs and outputs

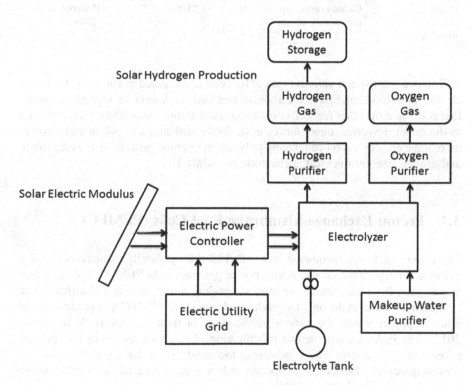

Fig. 3.2 The diagram of hydrogen production by electrolysis (Barbir 2005)

Table 3.1 Summary of electrolyte materials, transported irons and operating temperature of common fuel cells

Type of fuel cell	Material of electrolyte	Ions transfer	Environmental temperature (°C)	References
Solid oxide	Conducting ceramic oxide	O^{2-}	600–1000	Shao and Haile (2004)
Molten carbonate	Molten alkaline carbonate	CO_3^{2-}	600–700	Gemmen et al. (2000)
Alkaline	Aqueous alkaline solution	H^+	<250	Bimboim and Doly (1979)
Phosphoric acid	Molten phosphoric acid	OH^-, Na^+	150–200	Munson (1964)
Direct borohydride	Aqueous alkaline solution	H^+	20–80	Gyenge (2004)
Direct formic acid	Cation conducting polymer membrane	H^+	20–80	Rice et al. (2002)
Direct methanol	Cation conducting polymer membrane	H^+	20–80	Wasmus and Küver (1999)
Polymer electrolyte membrane	Cation conducting polymer membrane	H^+	20–80	Hamrock et al. (2008)

The characteristic of various types of fuel cell is discussed as shown in Table 3.1 above. Solid oxide and molten carbonate fuel cell are known as high temperature, fuel cell while the other fuel cells can be operated at low temperature. Fuel cell such as direct boronhydride, direct formic acid, direct methanol and polymer electrolyte membrane employ cation conducting polymer membrane material as the electrolyte and the ion transfer over to the electrode is mainly H^+.

3.3   Proton Exchange Membrane Fuel Cells (PEMFC)

The proton exchange membrane fuel cell (PEMFC) generally maneuvers on pure hydrogen fuel to tailor the efficacious power generation. In PEMFC, hydrogen fuel and oxygen from the atmosphere react to produce water, heat and electricity. The formation of water as the only by-product acknowledge PEMFC as a promising and efficient power source for various applications in industrial levels (Wang et al. 2013). The PEMFCs with neither modification of gas compressors or fuel pumps, gives a simpler system whilst increasing the conductivity for the performance of powering devices. Besides, they operate in low temperature, allows a better usage option in daily life (Li et al. 2004).

3.3.1 Properties of PEM Fuel Cell

the solid polymeric membrane is employed as a water-based electrolyte in PEM system for proton transfer. Protons are able to penetrate from ionic polymer electrolytes which does not conduct electrons in the electrolyte. The chemical reactions and the flow of electrical current are shown in Fig. 3.3.

The operation of fuel cell can be explained as: on the anode site, the hydrogen molecule is split into protons and electrons. The protons flow across the electrolyte to the cathode site while the electrons move through an external circuit. On the other side, the ambient air (contains oxygen) is pumped into the fuel cell on the cathode site. As depicted in the diagram above, the anode and cathode sites are connected by an electrical circuit whereby electrons move towards cathode site through the electrical circle. The protons and electrons will be combined with oxygen to form water at the cathode as by-products. The excess water goes out when the action runs regularly. The heat will be formed and will go out at the same time. The reactions inside the PEMFC are:

$$Anode: 2H_2 \rightarrow 4H^+ + 4e^- \tag{3.2}$$

$$Cathode: O_2 + 4H^+ + 4e^- \rightarrow 2H_2O \tag{3.3}$$

$$Overall: 2H_2 + O_2 \rightarrow 2H_2O \tag{3.4}$$

Fig. 3.3 The diagram of a proton exchange membrane fuel cell (Um et al. 2000)

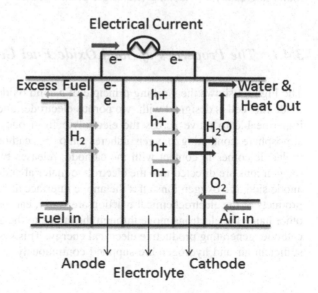

3.3.2 Advantages and Disadvantages of PEM Fuel Cell

The unique features of PEM to be made as the next promising and attractive fuel cell are being explored. This is because these fuel cells are able to be sustained at high current density, thus resulting in a fast start-up, compact and lightweight system. In addition, the low operating temperature (<100 °C), also aids in enhancing the rapid start-up. They are adopted by major auto makers. Moreover, the fuel cell is affordable in price and a long stack span life allows suitability for discontinuous operation (Tong et al. 2004). The PEM fuel cell has a potential efficiency at 50% of its performances. On the other hand, the disadvantage of the PEM fuel cell is the low operating temperature. Temperatures approximately or less than 100 °C are not high enough to perform effective cogeneration. In conjunction, catalyst (platinum) is usually required in the PEM cell as a promoter in the reaction to increase the performances of the operation. Heavy auxiliary equipment, and hydrogen at its purity are essential for PEM fuel cell (Deng et al. 2012).

3.4 Solid Oxide Fuel Cells (SOFC)

Solid oxide fuel cells utilize hard, non-porous metal oxide material as an electrolyte. In SOFCs, the negative oxygen ions from the cathode moves towards the anode through the motion of electrolyte. SOFC operates at elevated temperature ranging between 500 and 1000 °C thus usage of expensive catalyst such as platinum is not required in lowering the temperature fuel cells such as DMFCs. Thus, this eliminates proton exchange membrane between fuel cells and directly preventing carbon monoxide catalyst poisoning (Bettinger et al. 2008).

3.4.1 The Properties of Solid Oxide Fuel Cell

The Fig. 3.4 shows the working principle of a solid oxide fuel cell. Basically, solid oxide fuel cell is designed with two porous electrodes and an electrolyte. The gas is impermeable to move across the electrolyte from one electrode to another. The atmosphere containing oxygen adheres onto the cathode and when the oxygen molecule comes in contact with the cathode; releases two oxygen ions. Then, the oxygen ions are directed into the electrolyte material and move to the anode. At the anode side, the oxygen ions affix the anode interface and react catalytically. The by products of the electrochemical reaction are water, carbon dioxide and heat. On the other hand, the electrons move through the anode to the external circuit and back to cathode, generating productive electrical energy. This system is a repeating cycle if sufficient air and hydrogen are supplied continuously.

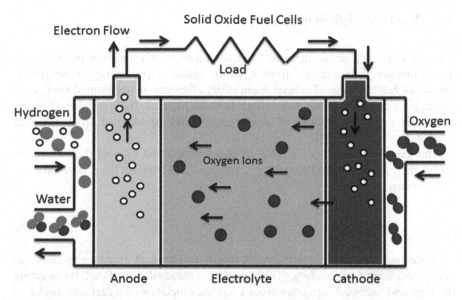

Fig. 3.4 Schematics of reactant flow in a SOFC (Slinn et al. 2008)

The reaction equation of SOFC

$$Anode: H_2 + O^{2-} \rightarrow H_2O + 2e^- \tag{3.5}$$

$$Cathode: O_2 + 4e^- \rightarrow 2O^{-2} \tag{3.6}$$

$$Overall: 2H_2 + O_2 \rightarrow 2H_2O \tag{3.7}$$

3.4.2 Advantages and Disadvantages of Solid Oxide Fuel Cell

One of the main highlights of solid oxide fuel cell is its high efficiency. Yet recent studies, efficiency of the solid oxide fuel cell is between 40 and 50%, and predicted efficiency of 60% does exists. Besides, solid oxide fuel cell employs the natural gas directly as well as a ceramic metal oxide electrolyte; thus, preventing occurence of corrosion and electrolyte movement or flooding in. The high operating temperature of solid oxide fuel cell can be used in achieving internal reforming, which convert into effective current energy. However, there are some drawbacks of this system. In accordance to its elevated temperature, there are several constraints on the material options as in the production processes due to most materials are unable to withstand such high temperature (1000 °C). Moreover, the high production cost of electrical energy from solid oxide fuel cell is a huge issue.

3.5 Fuel Cell Efficiency, η

Fuel cell efficiency can be described as the ratio of electrical energy produced for the production of heat gained from a fuel by burning. The energy of the fuel is known as heating value. The maximum of its efficiency is not limited by Carnot cycle, but occurs when the cell voltage is at its highest. These efficiencies have concentrated on the amount of power that could be attained from the fuel (Hirschenhofer et al. 2000).

The thermal efficiency of a fuel cell conversion is the ratio of the amount of useful energy produced to the change in enthalpy, ΔH between the product and the feed streams.

$$\eta = \frac{E_{useful}}{\delta H} \tag{3.8}$$

Since fuel cells are instruments that transform chemical energy into electrical energy, δG is the free energy for Gibbs energy in the ideal case of fuel. By equating the First and Second Thermodynamics Law, the efficiency of a fuel cell model is:

$$\eta_T = \frac{\delta G}{\delta H} \tag{3.9}$$

From the fundamental principle of efficiency:

$$\eta = \frac{W}{Q_{in}} \tag{3.10}$$

whereby Q_{in} is the enthalpy of formation of the reaction taking place and W is the output energy. Since both the values can often be evaluated depending on the state of the reactant, the larger one of the two values (Higher Heating Value) is used (HHV). N is the number of moles of electrons and F is the Faraday constant.

$$\eta = \frac{\delta G}{HHV} = \frac{NFE}{HHV} \tag{3.11}$$

The maximum efficiency materializes when in an open circuit environment when the highest voltage is gained.

$$\eta_{max} = \frac{\delta G}{HHV} = \frac{NFG}{HHV} \tag{3.12}$$

For instance, at standard conditions of 0 °C and 1 atmosphere, the thermal energy in the hydrogen/oxygen fuel cell reaction is 285.8 kJ/mole, and the Gibbs energy is 237.1 kJ/mole. Therefore, from the calculation, the maximum efficiency of the fuel cell at the standard conditions is:

$$\eta_{ideal} = \frac{237.1}{285.8} = 0.83 \tag{3.13}$$

Materials supported by catalyst display great impact on the cost, durability, and performance of polymer electrolyte membrane (PEM) fuel cells. The highlight of this article outline a few crucial kinds of novel support materials for PEM fuel cells (including direct methanol fuel cells). They are nanostructured carbon materials, conductive doped diamonds and nanodiamonds, conductive oxides and carbides.

3.5.1 Support Materials for PEMFC

Proton exchange membrane (PEM) fuel cells is a device that employs electrochemical reaction into green and effective electrical energy. PEMFC has been used in many applications such as in vehicles and for stationary and portable applications; thus, the contribution in scrutinizing methods for PEM fuel cells has achieved significant progress in the last few decades (Ball and Wietschel 2009). Nonetheless, PEM fuel cells, as well as direct methanol fuel cells (DMFCs), are still outlying from the market launch, which is impeded by two crucial problems (Wang 2004; Lee et al. 2013; Larson and Keach). They are the over pricey cost of production and its poor reliability and short life span. Therefore, electrocatalyst support is important as it greatly influences the performances and the durability of PEM fuel cells (Shao et al. 2009). Most of the catalysts used today are Pt-based and is usually supported on porous conductive materials with a high specific surface area, which improved the dispersion of catalytic metals. The electrocatalyst support also functions by providing a high dispersion and a narrow distribution of Pt and Pt-alloy nanoparticles, which aids in the performance of the catalytic catalysts. Besides, the addition of support materials to the PEM cell can greatly influence the catalytic activity and the durability. The most well-known electrocatalyst support is Vulcan XC-72 carbon black. In addition, in recent years, emphasis on electrocatalytic metals have increased tremendously (Shelef and McCabe 2000; Wang and Su 2014) in developing novel catalyst supports, including novel nanostructured carbons, (Shao et al. 2009) carbon nano-fibers (CNFs) (Guo et al. 2008) and mesoporous carbon (Zhang et al. 2010), oxides, (Serp and Figueiredo 2009) carbides, (Fang et al. 2012) nitrides, etc. These support materials can be classified into two types. The primary supports, such as novel nanostructured carbons and conductive diamonds, and the secondary supports, such as oxides, which are primarily employed to alter and enhance the primary supports. However, both types can also be used as independent supports.

3.5.1.1   Carbon Nanotubes and Carbon Nanofibers

Novel carbon materials like carbon nanotubes (CNT) and carbon nanofibres (CNF) (CNTs8–10, 18, 19 and CNFs20) have been gaining popularity as a support for catalyst in PEM fuel cells due to their distinct structure and characteristics. Novel nanostructured carbons have advantages such as its distinct structure and electrical properties which enable high electrical conductivity and a uniquely particular interaction between catalytic metals and the CNT supports, carbon materials have lesser impurities as compared to carbon black which comprised of significant amount of organosulfur impurities leading to poisoning of Pt metal, carbon nanostructured materials are free from deep cracks that exist in carbon black whereby will hamper the catalytic activity due to the electrochemical triple-phase boundary on the deposited Pt nanoparticles, and lastly, the ability to assemble an ordered catalyst layer in PEM fuel cells, provide ease of mass/electron transport, resulting in higher cell performance (Zhang et al. 2013).

A report by Wu et al. and Tang et al. on comparison between SWNT and DWNTs-supported Pt catalysts toward methanol oxidation resulted in a promising performance in SWNT than those supported by DWNTs. This is due to the higher specific surface area of SWNT which allows better enhancement of dispersion of Pt nanoparticles (Tang et al. 2009). Besides, a sustainable graphic crystallinity of SWNT provides a better charge transfer at the electrode interface as well as lower bulk resistance furnished to the increased in catalytic performances of SWNT-supported Pt. However, controversial of the properties of SWNTs have been reported, for instance, Li et al (Baughman et al. 2002) scrutinized the catalytic activity of PtRu supported on a series of carbon nanotubes (single, double and multiple-walled) and on carbon black. The results showed that DWN-supported PtRu exhibit highest specific activity in an oxidation reaction in methanol whilst the activity as an anode catalyst were excellent in DMFC single cell tests. Another report on double walled nanotubes supported Pt also resulted an excellent catalytic performance towards the oxygen reduction reaction. The differences in results may due to the fabrication methods of catalyst. Even so, the modification of CNTs in the catalytic performances as a supported material has gained popularity and been studied (Gao et al. 2012; Chen et al. 2014) compared to carbon black (Vulcan XC-72).

To start it off, SWNT is a semiconductor and possess low electronic conductivity, which greatly affects the catalytic performances (Yang et al. 2012). For instance, bamboo-structured MWNTs was noted to display faster electron transfer on the electrode in the electrochemical processes (Shao et al. 2009) as compared to hollow structure MWNT. Bamboo-structured MWNT comprised of edge plane like defect sites in which the axis of graphite is at an angle to the axis of the nanotubes whereas hollow structured MWNT possess axis of graphite planes parallel to axis of the nanotubes. In addition, CNF supported nanocatalyst also shows improvement in its catalytic activity. According to a report (Basri et al. 2010), the oxidation reduction reaction of a platelet CNF-supported Pd are more responding as compared to fish-bone CNF-supported Pd where it dictated a higher ratio of edge atoms to basal atoms of p-CNF. These results are in agreement with Tsuji et al.'s which studied on PtRu/CNF catalyst with platelet CNF, which showed high performances towards methanol oxidation than tubular and herringbone CNFs. On the other hand, Steigerwalt et al. discovered that PtRu/graphitic carbon nanofibre nanostructured with tubular herringbone graphitic CNF depicted a higher activity in comparative to single-, multiple-, and wide herringbone- graphitic supported PtRu catalyst. However, this has not been confirmed by an independent experiment. In addition, some suggested that twisted CNFs have better performances as compared to straight CNFs as a support in PEM fuel cell. Not only that, several opinions on smaller diameter of CNTs and CNFs shows better performance than others (Lee et al. 2006). Thus it can be concluded that structure defects in the supported catalyst does play a significant role in both providing ease of electron transfer as well as a guide in fashioning a PEM fuel cell with promising catalytic activity (Qin et al. 2012).

Another matter to be look into account is the durability of supports and catalyst. Carbon support which is widely used can be oxidized into carbon dioxide or carbon monoxide or any related oxygen species which is harmful and poisoning to the internal PEM fuel cell. Moreover, another catalytic metal supports such as Pt will disengage from the support, lose its performances and worst, the interactions between carbon and catalyst will be weakened. The resistance to electrochemical corrosion of raw MWNTs has recently been compared with that of Vulcan XC-72 carbon black under simulated PEM fuel cell conditions using the accelerated degradation test (ADT) techniques (Wang et al. 2008) For example, CNT and carbon black electrode are degraded by diluting it in an acid solution at high potential and this showed an increase in surface oxygen content on the Vulcan carbon electrode than for MWNTs, denoting CNTs are more resistance to electrochemical corrosion (Wu and Xu 2007) The durability of MWNTs also affected on their diameters, whereby medium-size MWNTs displaying better stability under electrochemical stressing. (Antolini 2009) Besides, under the ADT conditions CNT-supported Pt catalyst (Pt/CNT) was discovered to be more durable than the carbon black supported Pt/C (He et al. 2011).

Thus, to further enhance the durability of Pt/CNTs, a strong corrosion of CNTs, ability of catalyst to interact on the specific site of the CNTs, and allows high-temperature graphitization of CNTs process (Takenaka et al. 2007). It has been reported that the durability of Pt/MWNT is better than that of Pt/SWNT (Wu and Xu 2007). Meanwhile, Pt/DWNT has a slightly higher stability compared to Pt/MWNTs. (Chen et al. 2007). Conclusively, all CNTs supported Pt showed better durability compared to Pt/carbon. Therefore, novel nanostructured carbon materials (CNT/CNFs) do have potential in designing a promising PEM fuel cell as it displayed good performances in both catalytic activity and durability. However, improvement using a properly nanostructured carbon and new synthesis methods for carbon materials with controlled nanostructures and catalytic performances needs to be developed and established.

3.5.1.2   Mesoporous Carbon

Mesoporous carbon can be categorized into two types according to their structures and morphologies (Pérez-Ramírez et al. 2008). They are ordered mesoporous carbon (OMC) or known as ordered mesoporous silica (OMS) templates and disordered mesoporous carbon (DOMC). OMC are basically fabricated by nanocasting or by directly templating triblock copolymer structure-directing species, whereas DOMC are usually irregular pore structures, isolated or interconnected. DOMC typically displayed low conductivity, and the dispersal of pore sizes, is comparatively wider than in OMC. Thus, this allows OMC as better catalyst support in terms of electrical conductivity, specific surface area, and mass transport.

In contrast to traditional carbon support, mesoporous carbon shows a unique property structure as a catalyst support, such as, 3D interconnects mesopores large surface area with mono-dispersed. Thus, this material has been studied extensively for PEM fuel cells. The mesoporous Pt alloy catalysts and carbon-supported Pt and even other metal catalysts displayed promising catalytic performances due to the novelty and evenly distributed catalytic metals, their excellent conductivity and ability of mass transfer across the pore structure (Liu et al. 2004). Thus, to enhanced the performances of PEM fuel cell, an effective reaction site in nanoscale is essential. To materialize this, an ordered mesoporous SBA-15 silica template was initially infiltrated with carbon using a chemical vapor deposition (CVD) technique alongside with benzene function as carbon precursor to obtain a carbon/SBA-15 silica composite. Subsequently, a metal nanoparticle catalyst was supported on the surface of the composite. Then, the growth of CNTs was carried out on the composite surface and lastly, the silica framework and metal catalyst were eliminated to form bridged CNT/OMC particles.

Another method is to alter the mesoporous carbon with polypyrrole (PPy). It has been demonstrated that OMC selectively covered the outer surface with PPy, (Salgado et al. 2010) showed lower value of electrical resistance as compared to that of pristine OMC. Nonetheless, when the amount of Ppy loading increases, the resistance increased to the value of pristine OMC. However, in long-term run, stability of PPy in an acid environment is substandard, thus this improvision of electrical conductivity of OMC need to be in consideration for application in PEM fuel cells. The "bridge strategy" provides a new alternative to enhanced the performances of electronic conductivity in PEM fuel cell. Primarily, carbons in nanoscale (CNT, CNF, and mesoporous carbon) are typically used as supports for PEM fuel cells as these carbons have a higher durability. However, several challenges such as the high cost of carbon nanotubes as compared to carbon black impedes the commercializing of nanostructured catalyst supports of PEM fuel cell. Alternatively, doping with a secondary element (such as nitrogen 16) and the nanostructured carbons composites, or other compounds seems promising and legit in aiding the better performances for PEM fuel cell, which is discussed below.

3.5.1.3 Conductive Diamonds

Doping of diamonds inclusive of secondary element, such as boron produces boron-doped diamond (BDD) which can be classified as conductive diamonds. These properties have been scrutinized for the past two decades as a potential electrode in most of the electrochemical fields due to their unique properties such as low current background, wide potential window and exceptionally high chemical and dimensional stability. Nowadays, durability of the materials in fabricating PEM cell has been a moot (Highfield et al.). Thus the introduction of conductive diamonds has been investigated and developed. (Lackner et al. 2010) Since Pt nanoparticles are typically not adhered to the inner micropores surface, it brings advantages for the catalyst support (Home; Appleby 1996). Besides, harsh treatment such as thermal decomposition enables enhancement in mechanical and chemical stability of doped diamond without any changes in properties on the surface of the substrate, (Luong et al. 2009) thus allowing more opportunities to generate the catalyst with new properties and structure. Lately, developing conductive diamond with doping as a support catalyst for PEM fuel cells has been gaining much attention (Zhou et al. 2010; Antolini 2009).

Wang and Swain conducted a study on Pt/BDD by incorporating Pt nanoparticles into the carbon matrix and results showed improved stability of the resultant catalyst when no loss in activity was noted after 2000 potential cycles between the oxygen and hydrogen evolution regimes, (Shao et al. 2009) with no loss of catalyst activity for hydrogen evolution or oxygen reduction was observed after harsh electrolysis in H_3PO_4 at 170 °C, and no changes in its morphologies. Even though Pt nanoparticles can be deposited by several means of techniques, its relatively large size is to big for the usage in PEM cells (Cavaliere et al. 2011). This causes accumulation of complete coverage of smaller particles and partial coverage of the base of larger particle where roughly about one-third active Pt surface is lost after the secondary diamond deposition. Therefore, improvement in the dispersion of Pt nanoparticles is essential and lessen the negative effect of the coverage of Pt particles from the secondary diamond deposition (Wang et al. 2015).

Nonetheless, BDD is denoted as a good substrate in investigating the intrinsic properties of deposited catalytic nanoparticles, (Antolini 2009) and helps preventing complication with other common substrates, for instance, oxide formation as this is advantageous in the study of basic principles of electrocatalysis. The highly stability of conductive boron-doped diamond also encouraged the study of the catalytic properties of noble-metal oxide in the alteration from the level of sub-monolayer to a few monolayers, preventing severe interference of the reactivity of the adopted metal supports (Rabis et al. 2012). On the other hand, complication such as low conductivity, low surface area, and poor dispersion of the catalytic metals of doped diamonds supported catalyst makes it hard to materialize in a homogeneous and controlled level in diamond powders. Pure diamond is well-known as an insulator; thus by introducing doping (Crespi et al. 1996) the number of mobile charge carriers and defects increased simultaneously with increased in stability. Moreover, incorporation of surface functional groups, for instance, developing a hybrid diamond-graphite nanowires alleviate in prompting good surface area and electrical conductivity. (Hu et al. 2006) Based on the factual experiment, conductive diamonds displayed excellent durability under electro-chemical stressing conditions compared to other support materials. Thus, it can be proposed that BDD work better for fuel cell cathodes as cathodes can withstand severe corrosive conditions as compared to anodes (Liang et al. 2013).

3.5.1.4   Conductive Oxides

Conductive or semi-conductive oxides have been investigated mainly as a catalyst support material or act as a secondary support in modifying and promoting the catalyst support for PEM fuel cell as it displayed potential influences on the cat-alytic performances (Allaoui et al. 2002). Exemplary of these oxides are indium tin oxide (ITO), TiOx, WOx, IrO_2, SnO_2, and etc. Pt catalyst supported on ITO showed a change in the reaction pathways from 2e- and 4e-partial oxidation to 12e-complete oxidation in the oxidation reaction of ethanol. In addition, Chang et al. scrutinized the electrocatalytic properties of Pt nanoparticles adhered to ITO (PtNP/ITO) in the oxygen reduction, and methanol oxidation reaction and Pd nanoparticles on ITO (PdNP/ITO) for oxygen reduction. The results displayed that the methanol oxidation behavior on PtNP/ ITO is almost equivalent to that on bulk Pt, with faster kinetics, and the oxygen reduction on PtNP/ITO and PdNP/ITO is the same to that on bulk Pt and Pd, further verify that ITO has indeed been an excellent support in PEM application. Since PEM cells tend to submit to corrosion, alteration of the support surface with semi-conducting oxides, like WO_3 or coating with carbon allows minimization of corrosion to occur (CIIT 2016).

3.5.1.5   Titanium Oxides (Ti_yO_x)

Titanium oxide is a semiconductor and possess good electrical conductivity with a band gap of 4.85 eV and has been used widely in photocatalysis (Cahen et al. 2000) and as an electrocatalyst support material. In addition, these materials also dis-played high electrochemical stability (Shao et al. 2009) and distinct proton con-ductivity (Aksoy and Caglar 2014) which is advantageous in the formation of the electrochemical TPB. With these characteristics, enhancement of the catalytic performances is promising in the fuel cells (Lenzmann et al. 2001). It has been reported that Pt/alloys and Pd adhered on these oxides have excellence performance toward oxidation reduction reaction and modulus of resilience. In the case of PtRu adhering on TiO_2 supported nanoporous towards methanol oxidation, it can be observed that an obvious occurence of CO-poisoning on the cluster $PtRu/TiO_2$ support film as compared to unsupported PtRu or bare Pt catalysts. However, when TiO_2 thin film is sandwiched between the Pt and the Nafion membrane, the fuel cell performance elevated than that of Pt film adhered directly on the Nafion membrane.

Thereby, influencing good dispersal of metal nanoparticles on TiO_2 whilst improving its performances (Subramanian et al. 2001).

Besides, another alternative in enhancing the performances towards oxidation and reduction is by doping titanium oxide with metal (Dhakshinamoorthy et al. 2012). The high dispersion of the metal nanoparticles encourages mobility between oxide support and metal catalyst, whereby Pt L edge gained from the support catalyst can be observed through the X-ray absorption and thus improving its catalytic activity. Garcia et al. studied on the activity of mass of $PtRu/Nb_{0.1}Ti_{0.9}O_2$ and PtRu/C toward the modulus of resilience and results showed that PtRu/C is lower. Therefore, it can be said that introduction of titanium oxides in PtRu improves the catalytic activity as compared to carbon. Several ways of fabrication such as hydration and one-pot synthesis can boost the performances, but the ultimate techniques were reduction methods where reducing H_2PtCl_6 on carbon supported hydrated titanium oxide (TiO_2/C) displayed the highest performance toward oxygen reduction in DMFC and high resistance towards alcohol oxidation. This technique has been vastly utilized in fabricating of Pt/TiO_2-CNT and $PtRu/TiO_2$-C in reaction ethanol/ methanol oxidation; which showed high catalytic performance. In conjunction, mixture of titanium oxide with carbon-supported catalysts can improve the activity. For instance, a report showed an addition of Au/TiO_2 to a PtRu/C electrode has been shown to improve the performance of DMFCs. Mixture of this gives benefits to higher utilization of Pt, higher tolerance towards the formation of CO and favorable towards ethanol oxidation. When electrocatalyst are favorable towards ethanol oxidation, it makes formation of hydroxide more easier, thus encourages ethanol to be trapped and this boost the local concentration of ethanol around the Pt (Carp et al. 2004).

Another profound investigation was the carbon-coated anatase titania (CCT) TiO_2@C core–shell-like structures investigated by Shanmugam et al. They utilize Pt nanoparticle supports in the oxidation reduction reaction and modulus of resilience and results showed the novel nanomaterials exhibit a higher catalytic activity and stability as compared to commercial Pt/C. This is because the CCT possesses higher conductivity, better dispersion along with stability and ease in the interaction on the oxide surface as compared to carbon black. Not only that, the tolerance of charge-transfer for methanol oxidation on Pt/CCT was approximately half that of commercial Pt/C. Newly developed titania nanotubes are also being utilized on Pt and Pd nanocatalyst supports, which displayed excellent activity towards modulus of resilience than commercial ETek Pt/C and Pd-TiO_2 nanoparticles (Tahir and Amin 2013).

3.5.1.6 Tungsten Oxides (WO_y)

Tungsten oxide is a semiconductor (n-type) and can be formed in various states as it possesses different oxidation states; allowing suitability in application of photocatalyst, photochromic, sensor for gas and electrochromic (Carp et al. 2004). The innate electrical properties of tungsten oxide come from its non-ideal ratio composition, enable a rise in the donor level constituted by vacancy defects of oxygen in the lattice, which brings to a band gap of 2.6–2.8 eV. Besides, with an excellence reported tolerance on carbon monoxide and hydrogen sulphide, enhanced electrooxidation of methanol and ethanol, and reduction of oxygen, this oxide has been utilized ever since in the Pt/alloy electrocatalyst. Tungsten oxide also displayed a promising performance in the proton transfer, which have been reported on PtRu-WO$_3$ nanomaterial electrodes for oxidation methanol reaction. This is due to the formation of tungsten trioxide hydrates, which act as a unique characteristic that is essential as a support in the fuel cell catalyst and can be employed as a catalyst supports by itself. Not only that, these oxides can materialize into a hydrogen bronze that potentially aids in the dehydrogenation of methanol (Gospodinova and Terlemezyan 1998).

Alongside, newly developed tungsten oxide in nanoscale has been investigated and reported for application in electrocatalyst fuel cell. As such, Maiyalagan et al scrutinized an experiment on Pt supported on tungsten oxide nanorods and his results showed promising catalytic performances and excellent stability towards the oxidation of methanol. He further concluded that the best performances was the synergistic effect between Pt and WO$_3$ that prevents the electrodes from poisoning. Another experiment found that altering carbon supported RuSex nanoparticles with super thin films of WO$_3$ can display a better catalytic activity characteristics towards oxidation reduction reaction in an acidic medium and a mixed-valent tungsten oxide support can be enhanced the electronic and proton transfer conductivities. This is when a double functional electrocatalytic mechanism on RuSex causes the reduction on electrocatalytic of oxygen, WO$_3$ aids in the putrefaction of unwanted hydrogen peroxide and effectively reduce the total reaction to a 4e-process. This novel research is essential in developing non-Pt electrocatalyst as usually Pt/WO$_3$ catalyst showed high stability, but tungsten dissolution as well as the chemical instability of WO$_3$ in acidic medium [153] was still in findings to improve the fuel cell catalyst (Yamazoe 1991; Grätzel 2001).

3.5.1.7 Carbides

Since 1970s, transition metal of carbides has been scrutinized, particularly tungsten carbides, as it possesses platinum like catalytic properties, ability to be used as electrocatalyst as itself, or as catalyst support, and have more flexibility. This distinct property makes carbides an interesting material to work with as it allows enhancement in its catalytic activity as well as the feasibility of modification of chemical composition and surface during fabrication or post treatment. Besides, it has been reported that tungsten carbides are the most favorable electrocatalyst as compared to other carbides because of their stability in acidic medium, their specific selectivity, and particularly tolerant towards CO, methanol, and water and lastly its good electrical conductivities.

Research has been developing and studying on tungsten carbide with electrocatalyst, and the work showed that indeed tungsten carbides can operate as electrocatalyst by itself as well as supported by catalyst with synergistic effect along with catalytic metals or Pt. Introduction of metals helps higher its specific activity of the mechanism. For instance, Pt/tungsten carbide was observed in the reaction in methanol oxidation, and its function at the anode catalyst for microbial fuel cells and reported amelioration of performance in oxidation reduction reaction compared to Pt/C. These propitious results are due to the uniform dispersal of Pt nanoparticles, attributing to its high electrochemical surface area. According to Nie et al., the synergistic effect of Pt and WC in oxidation reduction reaction of Pt/WC/C showed that WC are an excellent catalyst supports. Other investigation on tungsten carbide were also conducted, such, WC supported Pt-Ru catalyst displayed a high resistance of CO. Enhancing the catalytic performances are also being studied, whereby the supported catalyst needed to exhibit good durability under performing in a fuel cell (Gasteiger et al. 2005; Robertson et al. 2010).

Chhina et al. conducted an experiment by investigating the durability of Pt/WC under ADT protocol with different potential steps. Results showed that degradation in activity for Pt/WC was only about 10–20% after 30 oxidation cycles. This indicates that oxidation of WC formed substoichiometric WO_y which changes the mechanism of Pt supported WC to Pt/Wo$_y$ enable the encapsulation of WC core, thus increasing the Pt surface area and lowering the degradation rate in fuel cells; increasing its performances. However, no data on the durability of tungsten carbide supported Pt was observed. According to Zhang et al. reported that Pt/WC is likely to be stable at 1.2 eV which denotes a close approximately to PEM fuel cell conditions. In the context of durability, Zellner and Chen showed that the raw WC is much lower in value as compared to Chhina et al. probably due to the chemical structure. Since tungsten carbides depends on the chemical structure, this enables feasible methods in alteration of the properties such as introduction of secondary metals into the carbides lattice; making better the stability as well as the catalytic performances. In spite of that, there are some limitations of carbides, such as the elctrochemical surface of tungsten carbides supported Pt is low in the application of fuel cells. This is because when surface area is small, further optimization of Pt size is required. Thus, the operations depend largely on the catalyst in fulfilling a good dispersion of the particles (Desportes et al. 2005; Davis and Klabunde 1982).

3.6 Conclusion

In this advanced society, finding an efficient catalytic performance, better durability and low cost support materials for PEM fuel cells have been a hot topic. The current investigation on numerous nanostructured support catalyst for PEM cells displaying excellent performances were reviewed. These materials were such as discussed above, providing support to carbon materials to enhance the catalytic activity and durability. Some of the novel yet intriguing research has been done by 3M company. They formed a novel non-conductive nanostructured organic material on a highly-elongated Pt nanoparticles distributed to act as a conductive nanostructured thin film. The promising conductive nanostructured thin film were able to generate electronic conduction under stressed voltage cycling as well as high potential under PEM fuel cell environments. Besides, a non-conductive silica has been used as secondary support as a matrix to stabilize Pt nanoparticles. This shows that some non-conductive materials can be utilized as a support in the application for fuel cell; where employing Pt enhanced the performances and durability of the operation. Another worthy material is the carbons. It is vastly observed as a promising supported catalyst in its catalytic activity and durability (Wu et al. 2011).

However, despite the advantages, there are still plenty of amelioration needed to be done in developing revolutionary supported materials to enhance the catalytic performances. Thus, future work such as improving the fuel cell reliability and life span, its durability under various aggressive conditions, and at affordable range of cost should be emphasized. Besides, using the modern quantum chemistry calculations is scrutinized in search for better materials and studying their properties in terms of electrocatalyst and its support. For instance, mesoporous carbon nitride was initially predicted by quantum chemistry calculation which then being utilized as a potential fuel cell catalyst; resulted in promising performances. Quantum mechanical calculations were also employed in investigating the interaction between catalytic metals and support which are predicted to bring in-depth knowledge of understanding the acting mechanism of support materials to enhance the catalysts. In conclusion, the need to improvise the catalytic performances with betterment of support catalysts are still highly in debate. Therefore, studying new strategies and methods on the support catalyst is essential in forming a effective, promising fuel cell.

References

Aksoy S, Caglar Y (2014) Structural transformations of TiO 2 films with deposition temperature and electrical properties of nanostructure n-TiO 2/p-Si heterojunction diode. J Alloy Compd 613:330–337

Allaoui A, Bai S, Cheng H-M, Bai J (2002) Mechanical and electrical properties of a MWNT/epoxy composite. Compos Sci Technol 62(15):1993–1998

An T, Zhou Y, Liu G, Tian Z, Li J, Qiu H, Tong G (2007) Genetic diversity and phylogenetic analysis of glycoprotein 5 of PRRSV isolates in mainland China from 1996 to 2006: coexistence of two NA-subgenotypes with great diversity. Vet Microbiol 123(1):43–52

Antolini E (2009) Carbon supports for low-temperature fuel cell catalysts. Appl Catal B 88(1): 1–24

Appleby A (1996) Fuel cell technology: status and future prospects. Energy 21(7–8):521–653

Ball M, Wietschel M (2009) The future of hydrogen—opportunities and challenges. Int J Hydrog Energy 34(2):615–627

Barbir F (2005) PEM electrolysis for production of hydrogen from renewable energy sources. Sol Energy 78(5):661–669

Basri S, Kamarudin SK, Daud WRW, Yaakub Z (2010) Nanocatalyst for direct methanol fuel cell (DMFC). Int J Hydrog Energy 35(15):7957–7970

Baughman RH, Zakhidov AA, De Heer WA (2002) Carbon nanotubes—the route toward applications. Science 297(5582):787–792

Bettinger CJ, Zhang Z, Gerecht S, Borenstein JT, Langer R (2008) Enhancement of in vitro capillary tube formation by substrate nanotopography. Adv Mater 20(1):99–103

Bimboim H, Doly J (1979) A rapid alkaline extraction procedure for screening recombinant plasmid DNA. Nucleic Acids Res 7(6):1513–1523

Cahen D, Hodes G, Grätzel M, Guillemoles JF, Riess I (2000) Nature of photovoltaic action in dye-sensitized solar cells. J Phys Chem B 104(9):2053–2059

Carp O, Huisman CL, Reller A (2004) Photoinduced reactivity of titanium dioxide. Prog Solid State Chem 32(1):33–177

Cavaliere S, Subianto S, Savych I, Jones DJ, Rozière J (2011) Electrospinning: designed architectures for energy conversion and storage devices. Energy Environ Sci 4(12):4761–4785

Chen S, Duan J, Jaroniec M, Qiao SZ (2014) Nitrogen and oxygen dual-doped carbon hydrogel film as a substrate-free electrode for highly efficient oxygen evolution reaction. Adv Mater 26 (18):2925–2930

Chen Z, Deng W, Wang X, Yan Y (2007) Durability and activity study of single-walled, double-walled and multi-walled carbon nanotubes supported Pt catalyst for PEMFCs. ECS Trans 11(1):1289–1299

Ciit KT (2016) Photocatalytical degradation of Congo red (CR) dye via nano Titanium dioxide coated glass bead under UV light. COMSATS Institute of Information Technology, Lahore

Crespi VH, Benedict LX, Cohen ML, Louie SG (1996) Prediction of a pure-carbon planar covalent metal. Phys Rev B 53(20):R13303

Davis SC, Klabunde KJ (1982) Unsupported small metal particles: preparation, reactivity, and characterization. Chem Rev 82(2):153–208

Deng Y, Sun C-Q, Cao S-J, Lin T, Yuan S-S, Zhang H-B, Zhai S-L, Huang L, Shan T-L, Zheng H (2012) High prevalence of bovine viral diarrhea virus 1 in Chinese swine herds. Vet Microbiol 159(3):490–493

Desportes S, Steinmetz D, Hemati M, Philippot K, Chaudret B (2005) Production of supported asymmetric catalysts in a fluidised bed. Powder Technol 157(1):12–19

Dhakshinamoorthy A, Navalon S, Corma A, Garcia H (2012) Photocatalytic CO_2 reduction by TiO_2 and related titanium containing solids. Energy Environ Sci 5(11):9217–9233

Fang B, Kim JH, Kim M-S, Yu J-S (2012) Hierarchical nanostructured carbons with meso–macroporosity: design, characterization, and applications. Acc Chem Res 46(7):1397–1406

Gao C, Guo Z, Liu J-H, Huang X-J (2012) The new age of carbon nanotubes: an updated review of functionalized carbon nanotubes in electrochemical sensors. Nanoscale 4(6):1948–1963

Gasteiger HA, Kocha SS, Sompalli B, Wagner FT (2005) Activity benchmarks and requirements for Pt, Pt-alloy, and non-Pt oxygen reduction catalysts for PEMFCs. Appl Catal B 56(1):9–35

Gemmen RS, Liese E, Rivera JG, Jabbari F, Brouwer J (2000) Development of dynamic modeling tools for solid oxide and molten carbonate hybrid fuel cell gas turbine systems. In: ASME Turbo Expo 2000: Power for Land, Sea, and Air. American Society of Mechanical Engineers, pp V002T002A068-V002T002A068

Gospodinova N, Terlemezyan L (1998) Conducting polymers prepared by oxidative polymerization: polyaniline. Prog Polym Sci 23(8):1443–1484

Grätzel M (2001) Photoelectrochemical cells. Nature 414(6861):338–344

Guo YG, Hu JS, Wan LJ (2008) Nanostructured materials for electrochemical energy conversion and storage devices. Adv Mater 20(15):2878–2887

Gyenge E (2004) Electrooxidation of borohydride on platinum and gold electrodes: implications for direct borohydride fuel cells. Electrochim Acta 49(6):965–978

Hamrock SJ, Rivard LM, Moore GGI, Freemyer HT (2008) Polymer electrolyte membrane. Google Patents

He D, Zeng C, Xu C, Cheng N, Li H, Mu S, Pan M (2011) Polyaniline-functionalized carbon nanotube supported platinum catalysts. Langmuir 27(9):5582–5588

Highfield J, Yasuda K, Siroma Z, Ioroi T, Nishimura Y, Oguro K Innovative electrocatalyst development and ionomer membrane studies for PEM electrochemical applications at the Osaka National Research Institute

Hirschenhofer J, Stauffer D, Engleman R, Klett M (2000) Fuel cell handbook. Business/Technology Books, Home FH The Beaulieu Hydrogen Home

Hu Y, Shenderova OA, Hu Z, Padgett CW, Brenner DW (2006) Carbon nanostructures for advanced composites. Rep Prog Phys 69(6):1847

Lackner KS, Dahlgren E, Graves C, Meinrenken C, Socci T, Archer L, Banerjee S, Castaldi M, Elimelech M, Fthenakis V (2010) Closing the carbon cycle: Liquid fuels from air, water and sunshine. Lenfest Center for Sustainable Energy, The Earth Institute, Columbia University, New York

Larson A, Keach S United Technologies Corporation Fuel Cells: Innovation Inside a Large Firm

Lee K, Zhang J, Wang H, Wilkinson DP (2006) Progress in the synthesis of carbon nanotube-and nanofiber-supported Pt electrocatalysts for PEM fuel cell catalysis. J Appl Electrochem 36 (5):507–522

Lee H-J, Kim J-H, Won J-H, Lim J-M, Hong YT, Lee S-Y (2013) Highly flexible, proton-conductive silicate glass electrolytes for medium-temperature/low-humidity proton exchange membrane fuel cells. ACS Appl Mater Interfaces 5(11):5034–5043

Lenzmann F, Krueger J, Burnside S, Brooks K, Grätzel M, Gal D, Rühle S, Cahen D (2001) Surface photovoltage spectroscopy of dye-sensitized solar cells with TiO_2, Nb_2O_5, and $SrTiO_3$ nanocrystalline photoanodes: Indication for electron injection from higher excited dye states. J Phys Chem B 105(27):6347–6352

Li F, Xu J, Dou Z-T, Huang Y-L (2004) Data mining-based credit evaluation for users of credit card. In: Proceedings of 2004 international conference on machine learning and cybernetics. IEEE, pp 2586–2591

Liang B, Cheng H-Y, Kong D-Y, Gao S-H, Sun F, Cui D, Kong F-Y, Zhou A-J, Liu W-Z, Ren N-Q (2013) Accelerated reduction of chlorinated nitroaromatic antibiotic chloramphenicol by biocathode. Environ Sci Technol 47(10):5353–5361

Liu Z, Ling XY, Su X, Lee JY (2004) Carbon-supported Pt and PtRu nanoparticles as catalysts for a direct methanol fuel cell. J Phys Chem B 108(24):8234–8240

Luong JH, Male KB, Glennon JD (2009) Boron-doped diamond electrode: synthesis, characterization, functionalization and analytical applications. Analyst 134(10):1965–1979

Munson RA (1964) Self-dissociative equilibria in molten phosphoric acid. J Phys Chem 68 (11):3374–3377

Pauniaho S-L, Salonen J, Helminen M, Heikinheimo O, Vettenranta K, Heikinheimo M (2014) Germ cell tumors in children and adolescents in Finland: trends over 1969–2008. Cancer Causes Control 25(10):1337–1341

Pérez-Ramírez J, Christensen CH, Egeblad K, Christensen CH, Groen JC (2008) Hierarchical zeolites: enhanced utilisation of microporous crystals in catalysis by advances in materials design. Chem Soc Rev 37(11):2530–2542

Qin Y-H, Jia Y-B, Jiang Y, Niu D-F, Zhang X-S, Zhou X-G, Niu L, Yuan W-K (2012) Controllable synthesis of carbon nanofiber supported Pd catalyst for formic acid electrooxidation. Int J Hydrog Energy 37(9):7373–7377

Rabis A, Rodriguez P, Schmidt TJ (2012) Electrocatalysis for polymer electrolyte fuel cells: recent achievements and future challenges. Acs Catal 2(5):864–890

Rice C, Ha S, Masel R, Waszczuk P, Wieckowski A, Barnard T (2002) Direct formic acid fuel cells. J Power Sources 111(1):83–89

Robertson NJ, Kostalik HA IV, Clark TJ, Mutolo PF, Abruña HD, Coates GW (2010) Tunable high performance cross-linked alkaline anion exchange membranes for fuel cell applications. J Am Chem Soc 132(10):3400–3404

Salgado J, Alcaide F, Álvarez G, Calvillo L, Lázaro M, Pastor E (2010) Pt–Ru electrocatalysts supported on ordered mesoporous carbon for direct methanol fuel cell. J Power Sources 195 (13):4022–4029

Serp P, Figueiredo JL (2009) Carbon materials for catalysis. Wiley, Hoboken

Shao Y, Liu J, Wang Y, Lin Y (2009) Novel catalyst support materials for PEM fuel cells: current status and future prospects. J Mater Chem 19(1):46–59

Shao Z, Haile SM (2004) A high-performance cathode for the next generation of solid-oxide fuel cells. Nature 431(7005):170–173

Shelef M, McCabe RW (2000) Twenty-five years after introduction of automotive catalysts: what next? Catal Today 62(1):35–50

Slinn M, Kendall K, Mallon C, Andrews J (2008) Steam reforming of biodiesel by-product to make renewable hydrogen. Biores Technol 99(13):5851–5858

Subramanian V, Wolf E, Kamat PV (2001) Semiconductor—metal composite nanostructures. To what extent do metal nanoparticles improve the photocatalytic activity of TiO_2 films? J Phys Chem B 105(46):11439–11446

Tahir M, Amin NS (2013) Advances in visible light responsive titanium oxide-based photocatalysts for CO_2 conversion to hydrocarbon fuels. Energy Convers Manag 76:194–214

Takenaka S, Matsumori H, Nakagawa K, Matsune H, Tanabe E, Kishida M (2007) Improvement in the durability of Pt electrocatalysts by coverage with silica layers. J Phys Chem C 111 (42):15133–15136

Tang S, Sun G, Sun S, Qi J, Xin Q, Haarberg GM (2009) Double-walled carbon nanotubes as an electrode for direct methanol fuel cell applications. ECS Trans 16(50):113–122

Tong Z-B, Gold L, De Pol A, Vanevski K, Dorward H, Sena P, Palumbo C, Bondy CA, Nelson LM (2004) Developmental expression and subcellular localization of mouse MATER, an oocyte-specific protein essential for early development. Endocrinology 145(3):1427–1434

Um S, Wang CY, Chen K (2000) Computational fluid dynamics modeling of proton exchange membrane fuel cells. J Electrochem Soc 147(12):4485–4493

Wang C-Y (2004) Fundamental models for fuel cell engineering. Chem Rev 104(10):4727–4766

Wang D-W, Su D (2014) Heterogeneous nanocarbon materials for oxygen reduction reaction. Energy Environ Sci 7(2):576–591

Wang F, Bing Z, Zhang Y, Ao B, Zhang S, Ye C, He J, Ding N, Ye W, Xiong J (2013) Quantitative proteomic analysis for radiation-induced cell cycle suspension in 92-1 melanoma cell line. J Radiat Res rrt010

Wang J, Yin G, Shao Y, Wang Z, Gao Y (2008) Investigation of further improvement of platinum catalyst durability with highly graphitized carbon nanotubes support. J Phys Chem C 112 (15):5784–5789

Wang Y-J, Zhao N, Fang B, Li H, Bi XT, Wang H (2015) Carbon-supported Pt-based alloy electrocatalysts for the oxygen reduction reaction in polymer electrolyte membrane fuel cells: particle size, shape, and composition manipulation and their impact to activity. Chem Rev 115 (9):3433–3467

Wasmus S, Küver A (1999) Methanol oxidation and direct methanol fuel cells: a selective review. J Electroanal Chem 461(1):14–31

Wu G, Xu B-Q (2007) Carbon nanotube supported Pt electrodes for methanol oxidation: a comparison between multi-and single-walled carbon nanotubes. J Power Sources 174(1):148–158

Wu G, Johnston CM, Mack NH, Artyushkova K, Ferrandon M, Nelson M, Lezama-Pacheco JS, Conradson SD, More KL, Myers DJ (2011) Synthesis–structure–performance correlation for polyaniline–Me–C non-precious metal cathode catalysts for oxygen reduction in fuel cells. J Mater Chem 21(30):11392–11405

Yamazoe N (1991) New approaches for improving semiconductor gas sensors. Sens Actuators B Chem 5(1–4):7–19

Yang G, Han H, Li T, Du C (2012) Synthesis of nitrogen-doped porous graphitic carbons using nano-$CaCO_3$ as template, graphitization catalyst, and activating agent. Carbon 50(10):3753–3765

Zhang Y, Zhang X, Liu Z, Bian Y, Jiang J (2005) Structures and properties of 1, 8, 15, 22-tetra substituted phthalocyaninato-lead complexes: the substitutional effect study based on density functional theory calculations. J Phys Chem A 109(28):6363–6370

Zhang W, Sherrell P, Minett AI, Razal JM, Chen J (2010) Carbon nanotube architectures as catalyst supports for proton exchange membrane fuel cells. Energy Environ Sci 3(9):1286–1293

Zhang S, Shao Y, Yin G, Lin Y (2013) Recent progress in nanostructured electrocatalysts for PEM fuel cells. J Mater Chem A 1(15):4631–4641

Zhou Y, Neyerlin K, Olson TS, Pylypenko S, Bult J, Dinh HN, Gennett T, Shao Z, O'Hayre R (2010) Enhancement of Pt and Pt-alloy fuel cell catalyst activity and durability via nitrogen-modified carbon supports. Energy Environ Sci 3(10):1437–1446

Chapter 4
Hydrogen Energy in General

4.1 Introduction

Presently, finding a friendly, clean and efficient energy in generating fuel cells and hydrogen from various primary energy source has been the main solution. Skyscraper buildings and industrial sites, mostly generate heat, which accounts for more than one per-two in consumption of energy and one-third carbon dioxide and other pollutant gases. (Stambouli and Traversa 2002). Introduction of recent used hydrocarbon fuels in the application of heat generation along with low-carbon alternatives is said to reduce the greenhouse effect significantly by the year 2050 and prevent climate change. (Das and Veziroğlu 2001). It was then, in 2002, The High-Level Group for Hydrogen and Fuel Cells Technologies has begun to commence and was invited to fabricate and contribute in materializing a sustainable energy system in the near future. You can see that energy is the essence in the society and economies, such as in daily activities as well as economic, social and physical welfare largely influences by the supple, continuous supply of energy (Saxena et al. 2009).

The need for energy uninterruptedly grow year after year, however natural resources (i.e. fossil energy) is shrinking and limited in supply as the years goes by. In 2011, total global energy utilized for heat in buildings and industry was about 172 EJ (Stambouli 2011) and approximately 75% of the heat was formed by combustion of fossil fuels, thus piloting of 10 $GtCO_2$ emissions produced. Thus, alternative ways to overcome this issue is by introducing natural primary sources. Natural primary sources are naturally available and its such as from the wind, sun, coal and gas as well as geothermal. Attempt at using this technique provides a betterment in sustaining the energy whilst bringing positive impacts in the change of the global climate, decreasing the risk of supply disruptions, avoid rapid changes in price and lastly helps in reduction of air pollutions. In recent years, production of hydrogen has been gaining worldwide attention as a clean, efficient energy carrier that can be utilized in fossil fuels to generate power. With the combination of both

© Springer International Publishing AG 2017
S. Bagheri, *Catalysis for Green Energy and Technology*,
Green Energy and Technology, DOI 10.1007/978-3-319-43104-8_4

hydrogen and fossil fuels, hold excellent promises in securing our supply and climate change. As a result, in October 2002, the High-Level Group for Hydrogen and Fuel Cell Technologies have established a European Hydrogen and Fuel Cell Technology Partnership and Advisory Council to guide the process and denoted as the first milestone in developing novel, innovative methods for sustainable energy system. At the same time in May 2014, the UK Hydrogen and Fuel Cell Supergen Hub published a White Paper that systematically examines the evidence base in implementing hydrogen and fuel cells to generate low-carbon, secure, affordable heat in the UK (Larminie et al. 2003).

4.2 The Limitations

According to the European "World Energy Technology and Climate Policy Outlook" (WETO) presage the mean growth rate of 1.8% yearly for the period 2000–2030 for primary energy worldwide. The significant elevating demand for energy is being met mostly by the supply of fossil fuels that emanate greenhouse gasses and unwanted toxic pollutants, prior to be an alarming rate. Presently, level emission of carbon dioxide has increased a 20% per capita in developing countries from major industrial nations. As this industry rose, emission of CO_2 will substantially increase. It is predicted that by 2030, the CO_2 emissions in the developing countries will exceeded for more than half the world emissions of carbon dioxide. Thus, it is essential to regulate rules and policy to offset this issue.

Traditional resources such as fossil fuels, one of the main source of energy was incarcerated to a few areas of the world and the continuous supply was administered by political, economic and ecological aspect. This resulted in energy security problems such as the over-priced fuel rates and a persistent environmental policy in minimizing greenhouse effect and unwanted pollutants. Therefore, a sensible yet feasible technique is required in taking into account the production of fuel, transmission and distribution, the energy conversion and the impact on the equipment and the consumer when the energy system used. Nonetheless, the highlight is to develop a high achieving energy efficiency and elevated supply, typically renewable sources. Production of energy based hydrogen are being in consideration in the near future in technological developments as it provides both clean and efficient system. The High-Level Group emphasized the potential of hydrogen-based energy systems world-widely, especially in Europe, in terms of broad energy and environment plans.

4.3 Construction of Carbon Footmark

Fuel cells are large and bulky device that are produced by using catalyst metals such as nickel and platinum which are extremely energy-intensive. Typically, like the other low-carbon technologies, the energy needed to manufacture the fuel cell and the resulting carbon emissions is utmost crucial in operational work as they offset the savings made. Numerous life-cycle assessments (LCAs) have predicted the incorporated carbon or the carbon foot-mark by several factors such as the manufacturing of the fuel cell, materials produced and required in the fuel cell. Mass-production of a 1 kW residential CHP system results in 25–100 tCO_2, (Brentner et al. 2011; Guinee et al. 2010). This can be due to influence of the device used (e.g. between PEMFC and SOFC), but the majority is affected by the production and manufacturing techniques used by different brands. The carbon foot-mark would tremendously mitigate if the mass-production processes were decarbonized.

4.4   Why Hydrogen and Fuel Cells?

Supplying a clean, efficient, safe and secure energy are the main force in sustaining a high quality of life. However, energy systems must meet the demand of following societal needs in terms of cost, able to reduce the effects of climate change, mitigate toxic pollutants and able to eliminate reserves of oil. The adverse effect of this system nonetheless has a huge impact to the economy, environment as well as to the people itself. Thus, measures needed to be implemented in usage of more efficient, clean, green energy. Therefore, in today research, an ideal, emissions-free future based on sustainable energy such as electricity and hydrogen are scrutinized. The combination of electricity and hydrogen represent one of the most promising ways to achieve this, complemented by fuel cells which provide very efficient energy conversion. Hydrogen can be found easily and abundantly, which acts as an energy carrier. Hydrogen was first generated utilizing energy systems dependant on various renewable energy carriers and sources.

The need for bulk quantities of hydrogen in markets of transport as well as in stationary power could impede the progress overbroadly from the original demonstration phase. If the highlighted aim is in cost and security, then producing syngas with carbon dioxide sequestration can be of interest for large parts of Europe. Various ranges of options for sources, converters and applications are as depicted in Figs. 4.1 and 4.2, although not comprehensive, highlighted the flexibility of hydrogen and fuel cell energy systems. Fuel cells can be seen being utilized in a wide range of products, ranging from mini-sized fuel cells in portable devices to transport applications and also to heat and power generators in stationary applications in both domestic and industrial sectors. Future energy systems will also include enhanced traditional energy converters running on hydrogen along with other energy carriers.

The advantages of these systems are that it has little to almost carbon-free emissions and nontoxic as well as pollutant free substances such as nitrogen dioxide, sulphur dioxide or carbon monoxide. Besides, their low noise and high power quality, are ideal for use in hospitals or IT centers and also for mobile applications. Fuel cell system furnishes high efficiencies not depending on its size.

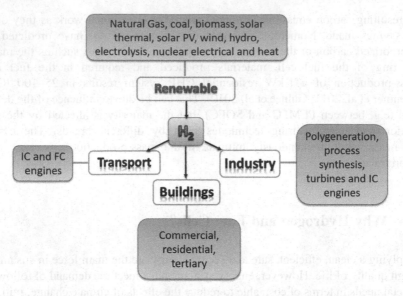

Fig. 4.1 Hydrogen: primary energy sources, energy converters and applications

Electric-drive trains based fuel cell can supply an excellent mitigation in energy consumption and regulated emissions. Auxiliary Power Units (APU) in combination with internal combustion engines, or in stationary back-up systems can incorporate fuel cell when operated with reformers for on-board conversion of other fuels. This largely increased the saving energy and alleviate air pollution. In summary, hydrogen and fuel cells open the way to integrate "open energy systems" that address all of the major energy and environmental challenges, and have the flexibility to revamp to the diverse and sporadic primary energy sources that will be abundant in the Europe of 2030.

4.5 Fuel Cells

Fuel cells are versatile, feasible device that are easily designed to be incorporated into private houses to large office blocks and industrial sites. They serve as the cogeneration system, in which provides the highest proportion of electricity consumption world widely. This system harness the production of heat and power whilst supplying tremendous high efficiency (95%), reducing its dependence on main power generator and help save the electricity cost as well as alleviate the carbon emissions. Besides, the fuel cell is the most renowned due to their supremacy in electrical efficiency.

Fig. 4.2 Fuel cell technologies, possible fuels and applications

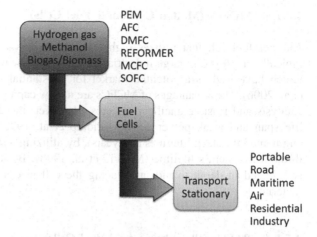

4.5.1 Types of Fuel Cells

4.5.1.1 PEMFCs (Proton Exchange Membrane Fuel Cells)

PEMFCs are commonly utilized and investigated, as it is able to supply around 90% of power shipped to date (Wang et al. 2004). Besides, Japanese 'EneFarm' program has shown that these cells can be employed in residential heating systems (1–3 kW thermal) and stack technology used in fuel cell vehicles. Years of effort in exploiting and researching of PEM become fruitful when these devices furnish with excellent efficiency, durability, reliability, as well as the affordable price. However, system complication is a challenge. In todays' research, eliminating the platinum to prevent complex engineering solutions (Yang et al. 2001; Song and Tsiakaras 2006), and operating at high temperature conditions to exclude the need for humidifiers (Um et al. 2000; Kim et al. 1995) are done.

4.5.1.2 SOFCs (Solid Oxide Fuel Cells)

SOFCs are fuel cells comprised of high-temperature utilize in large industrial and housing heating systems, that are currently expanding to reach 10% of global sales (Smith et al. 2009). Properties of SOFC include excellent electrical efficiency and betterment in versatility, however, these cells operate less dynamically compared to PEMFC because of the prerequisite temperature (Shao and Haile 2004). Another con is the sensitive operational system in starting and shutting which consume >12 h. This makes the system often hot, thus lowering the output efficiency. Therefore, investigation on augmenting its durability and material fatigue as well as operating the system at 500–750 °C (Minh 2004) are being studied. This ensures a bigger range of materials to be utilized, lowering its costs and enhancing the dynamic performance.

4.5.1.3   MCFCs (Molten Carbonate Fuel Cells)

Another fuel cell that utilizes high temperature is the MCFCs. These cells are typically employed in large industrial CHP and grid-scale electricity production, in which have lead as a potential market for substantial stationary applications (Li et al. 2006). The advantages of MCFCs are its low capital costs due to non-platinum catalysts and feasible ancillary systems. However, these system suffer from short life span and weak power density (Plomp et al. 1992) The main challenge has augmented the stack lifetimes (>5 years), by utilizing a stack replacement half-way through a system's lifetime (Maggio et al. 1998). Besides, power density is also scrutinized in depth to aid in reducing the cell size and thus lower cost of the material.

4.5.1.4   PAFCs (Phosphoric Acid Fuel Cells)

PAFCs were the first fuel cell device used for heating in the 70 s in commercial-scale CHP systems (Blomen and Mugerwa 2013). Approximately 400 systems are in operation, mainly in the Germany, US, Korea and Japan (Sammes et al. 2004; Li et al. 2006). These systems have been demonstrated in a small number on the scale of 1 kW (Watanabe et al. 1994), but no residential products have been commercialized.

4.6   Energy Security and Its Supply

Presently and in future to come, dependent on the continuous supply of cheap fossil fuels can lead to a major upturn in geopolitical as well in the price market. Implementing hydrogen as a primary energy sources, including nuclear energy, fossil fuels and renewable energy sources as an open access makes it more commonly obtained. With the ease in obtaining hydrogen, stability of the price of hydrogen as a carrier is maintained and enable exploitation of electricity resources. The combination of these two allow flexibility in maintaining equal power between centralized and decentralized locations. Decentralized power is attractive in providing the quality of power to meet specific customer needs and act as a reducing exposure to terrorist attack. Besides, its potential to store hydrogen with ease as compared to electricity aid in load levelling and balancing fluctuated power sources. Hydrogen is among the few energy carriers than providing renewable energy sources to be added into its transport system.

4.6.1   Economic Competitiveness

After the dispute of the oil crisis in 1970s, the growth of economic has not been dependent on the rise in energy demand in industrial sites, but quite the opposite for the transport sector. Therefore, it is of utmost important to lower the consumption of energy per unit growth in the development of energy carriers and technologies in order to create affordable energy supply. Growth and sales of these energies also contribute to the wealth creations fashioning substantial chances for employment and export, typically in the industrial sector. European leadership in hydrogen and fuel cells will play a crucial part in fashioning high-quality employment opportunities, in terms of R&D to production and craftsmen. There is robust investment and industrial activity in the usage of hydrogen and fuel cell sector in these countries, leading the development in hydrogen spreading independently in Europe.

4.6.2   Improvement of Health and Air Quality

Modified and development of traditional technologies and post combustion treatments for improved properties are without interruption aiding in lower emission of pollutants. However, in some area, oxides of nitrogen and particulates still persist. Thus, the need for green and clean energy solutions have to be ameliorated in public transport as the global trend towards urbanization accentuate it. With this, utilization of vehicles and power plant generated by hydrogen produces zero emission, thus benefits local air quality.

4.6.3   Alleviation of Greenhouse Gas

Production of hydrogen from carbon-free or fossil fuels are generated as carbon dioxide are captured and stored. By implementing this system (hydrogen and fuel cells), removal of greenhouse emission and efficient supply and clean electricity generation from a range of fuels can be ensured. Hydrogen can be produced from carbon-free or carbon-neutral energy sources or from fossil fuels with CO_2 capture and storage (sequestration). Thus, the use of hydrogen could eventually eliminate greenhouse gas emissions from the energy sector. Fuel cells provide efficient and clean electricity generation from a range of fuels. Besides, this system can also be sited for the ultimate application in which a product has been designed, harnessing the heat generated in the process. For instance, in 2008, the introduction of hydrogen fuelled vehicles in Europe help alleviate the mean emission of greenhouse gases as compared to the average level of 140 g/km CO_2, thus boosting the economy.

4.7 Hydrogen

An alternative to natural gas that is constantly used is the hydrogen. Hydrogen can be used as gas cooking, in water and space heating and other appliances by utilizing the Wobbe index comparison metric technique. In some Europe countries (Momirlan and Veziroglu 2005), hydrogen with high purity has a Wobbe index of approximately 48 MJ/m^3 (Olah 2005), which is within the natural gas safety regulation range for burners. However, even though hydrogen is in close range to others natural gas, it cannot be used directly in gas appliances. This is due to the combustion velocity (known as flame speed) which is higher as compared to natural gas. Thus, need to control the flame is required for various burner head designs (Ni et al. 2006). It is intelligible that the physical and chemical properties of hydrogen as well as the safety yardstick are crucial for industrial processes associated with hydrogen such as a fuel in buildings (Leung et al. 2010).

4.7.1 Hydrogen Heat Technologies

This technology has been incorporated in homes as a power fuel cell micro-CHP, direct flame combustion boilers (similar to existing natural gas boilers), catalytic boilers and gas-powered heat pumps. Besides, various bulky district heat and CHP machines in industries (Steele and Heinzel 2001) that utilized natural gas can be remodeled for hydrogen use (Acres 2001). In Europe and North America, direct flame combustion of hydrogen boiler function identified as a gas boiler providing to residential central heating (Ren and Gao 2010). When the catalytic boiler passes H_2 over a highly reactive metal catalyst, the exothermic chemical reaction is produce resulting heat for space and hot water heating without burning of flame. The operation displayed extremely low nitrogen oxide emissions, with feasible operation on the heat output (Steele 2001; Hawkes and Leach 2005). From a consumer perspective, catalytic hydrogen boilers are no different than the natural gas boilers and can be modeled for attractive and betterment in its performances. Subsequently, pump heat employing gas has been commercialized in the German market (Choudhury et al. 2013) especially in for larger commercial buildings and residential-scale models. This technology utilizes phase-change working fluid to absorb heat from an ambient source and to transfer it to the building heating system, increasing its thermal efficiency. Employing gases such as hydrogen help lessen emission of carbon dioxide and therefore reducing emissions across the energy system.

4.7.2 Production of Hydrogen

Back in the years, hydrogen has been utilized to produce ammonia as crop fertilizer and for "cracking" heavy oil into fuels such as kerosene, diesel, and petrol. It is also used in a various of industries such as food processing and metal fabrication (Ghasemi et al. 2012). Production of hydrogen is generated from biological material, fossil fuels or water (Dodds et al. 2015). Besides, few other techniques in development that could play a part in future hydrogen supply chains, including:

(a) Electrolysis at high temperatures, using heat from nuclear reactors or concentrating solar power (Steinfeld 2005), ensuring more efficiency in the process.
(b) Thermolysis, which employ extreme heat from nuclear or solar energy to split hydrogen from water (Abanades and Flamant 2006).
(c) Photocatalytic water splitting, the process of obtaining hydrogen directly from water using sunlight (Bard and Fox 1995).
(d) Hydrogen production from direct fermentation of biological material (Das and Veziroğlu 2001).

Unfortunately, hydrogen energy is not ideal as sustain energy sources, only if the emission of hydrogen production is relatively low. For instance, in Germany, a renewable hydrogen standard has been instigated by TÜV-SÜD, and is enormously employed as a benchmark in projects and demonstration activities for defining 'green' hydrogen. In addition, the European Commission Joint Undertaking on Hydrogen and Fuel Cells is endowing in the research of a European scheme in ensuring the origin of green hydrogen, in order for a harmonized development of green hydrogen standards.

4.7.3 The Advantages of Hydrogen Fuel Cells

(a) It is abundant—Can be easily obtained from the environment.
(b) Non-toxic—Eco-friendly towards the environment, unlike gasoline and petrol.
(c) Environment friendly—Burned hydrogen does not emit harmful gases such as carbon dioxide and other pollutants.
(d) Fuel in rockets—Hydrogen is efficient and powerful enough to supply energy to large machinery as it has three times as powerful as gasoline and other fossil fuels.

4.8 Pathways of Heat Decarburization

Pathways of decoking are usually fashioned by commodity flows representing the whole economy and used to associate with the energy system that meets energy service demands with the lowest discounted capital, operating and resource cost. These systems typically used to inform climate policy in many countries. Thus, it is crucial in fashioning an apt representation of hydrogen fuel cell heating devices in order for a betterment in the devices as well as the depth understanding liaising to policymakers (Linnen 2014).

However, as these models have an inclusive illustration of the whole energy system, they are likely to be focused on separated parts/sectors as well as coarse spatial and temporal resolutions, to provide a little complication in designing the model complexity and lessen the time taken for the computational interpretion time. Nonetheless of the compromised element, limitation in obtaining specific information about the speculate assumptions in such models are harder if the documentation is poorly or not made available (Ferrell 2014). Even though materializing of these targets is by no means guaranteed, however such that of micro-CHP which remain the lowest emission option could be predicted in the future.

References

Abanades S, Flamant G (2006) Thermochemical hydrogen production from a two-step solar-driven water-splitting cycle based on cerium oxides. Sol Energy 80(12):1611–1623

Acres GJ (2001) Recent advances in fuel cell technology and its applications. J Power Sources 100 (1):60–66

Bard AJ, Fox MA (1995) Artificial photosynthesis: solar splitting of water to hydrogen and oxygen. Acc Chem Res 28(3):141–145

Blomen LJ, Mugerwa MN (2013) Fuel cell systems. Springer Science & Business Media

Brentner LB, Eckelman MJ, Zimmerman JB (2011) Combinatorial life cycle assessment to inform process design of industrial production of algal biodiesel. Environ Sci Technol 45(16):7060–7067

Choudhury A, Chandra H, Arora A (2013) Application of solid oxide fuel cell technology for power generation—a review. Renew Sustain Energy Rev 20:430–442

Das D, Veziroğlu TN (2001) Hydrogen production by biological processes: a survey of literature. Int J Hydrogen Energy 26(1):13–28

Dodds PE, Staffell I, Hawkes AD, Li F, Grünewald P, McDowall W, Ekins P (2015) Hydrogen and fuel cell technologies for heating: a review. Int J Hydrogen Energy 40(5):2065–2083

Ferrell JC (2014) A distributed model of oilseed biorefining, via integrated industrial ecology exchanges. North Carolina Agricultural and Technical State University

Ghasemi Y, Rasoul-Amini S, Naseri A, Montazeri-Najafabady N, Mobasher M, Dabbagh F (2012) Microalgae biofuel potentials (Review). Appl Biochem Microbiol 48(2):126–144

Guinee JB, Heijungs R, Huppes G, Zamagni A, Masoni P, Buonamici R, Ekvall T, Rydberg T (2010) Life cycle assessment: past, present, and future. ACS Publications

Hawkes A, Leach M (2005) Impacts of temporal precision in optimisation modelling of micro-combined heat and power. Energy 30(10):1759–1779

Kim J, Lee SM, Srinivasan S, Chamberlin CE (1995) Modeling of proton exchange membrane fuel cell performance with an empirical equation. J Electrochem Soc 142(8):2670–2674

Larminie J, Dicks A, McDonald MS (2003) Fuel cell systems explained, vol 2. Wiley, Chichester

Leung DY, Wu X, Leung M (2010) A review on biodiesel production using catalyzed transesterification. Appl Energy 87(4):1083–1095

Li X, Fields L, Way G (2006) Principles of fuel cells. Platinum Metals Rev 50(4):200–201

Linnen MJ (2014) Advanced reactors and novel reactions for the conversion of triglyceride based oils into high quality renewable transportation fuels. The University of North Dakota

Maggio G, Freni S, Cavallaro S (1998) Light alcohols/methane fuelled molten carbonate fuel cells: a comparative study. J Power Sources 74(1):17–23

Minh NQ (2004) Solid oxide fuel cell technology—features and applications. Solid State Ionics 174(1):271–277

Momirlan M, Veziroglu TN (2005) The properties of hydrogen as fuel tomorrow in sustainable energy system for a cleaner planet. Int J Hydrogen Energy 30(7):795–802

Ni M, Leung DY, Leung MK, Sumathy K (2006) An overview of hydrogen production from biomass. Fuel Process Technol 87(5):461–472

Olah GA (2005) Beyond oil and gas: the methanol economy. Angew Chem Int Ed 44(18): 2636–2639

Plomp L, Veldhuis J, Sitters E, Van der Molen S (1992) Improvement of molten-carbonate fuel cell (MCFC) lifetime. J Power Sources 39(3):369–373

Ren H, Gao W (2010) Economic and environmental evaluation of micro CHP systems with different operating modes for residential buildings in Japan. Energy Build 42(6):853–861

Sammes N, Bove R, Stahl K (2004) Phosphoric acid fuel cells: fundamentals and applications. Curr Opin Solid State Mater Sci 8(5):372–378

Saxena R, Adhikari D, Goyal H (2009) Biomass-based energy fuel through biochemical routes: a review. Renew Sustain Energy Rev 13(1):167–178

Shao Z, Haile SM (2004) A high-performance cathode for the next generation of solid-oxide fuel cells. Nature 431(7005):170–173

Smith TR, Wood A, Birss VI (2009) Effect of hydrogen sulfide on the direct internal reforming of methane in solid oxide fuel cells. Appl Catal A Gen 354(1):1–7

Song S, Tsiakaras P (2006) Recent progress in direct ethanol proton exchange membrane fuel cells (DE-PEMFCs). Appl Catal B 63(3):187–193

Stambouli AB (2011) Fuel cells: The expectations for an environmental-friendly and sustainable source of energy. Renew Sustain Energy Rev 15(9):4507–4520

Stambouli AB, Traversa E (2002) Solid oxide fuel cells (SOFCs): a review of an environmentally clean and efficient source of energy. Renew Sustain Energy Rev 6(5):433–455

Steele B (2001) Material science and engineering: the enabling technology for the commercialisation of fuel cell systems. J Mater Sci 36(5):1053–1068

Steele BC, Heinzel A (2001) Materials for fuel-cell technologies. Nature 414(6861):345–352

Steinfeld A (2005) Solar thermochemical production of hydrogen—a review. Sol Energy 78(5):603–615

Um S, Wang CY, Chen K (2000) Computational fluid dynamics modeling of proton exchange membrane fuel cells. J Electrochem Soc 147(12):4485–4493

Wang C, Waje M, Wang X, Tang JM, Haddon RC, Yan Y (2004) Proton exchange membrane fuel cells with carbon nanotube based electrodes. Nano Lett 4(2):345–348

Watanabe M, Tsurumi K, Mizukami T, Nakamura T, Stonehart P (1994) Activity and stability of ordered and disordered co-pt alloys for phosphoric acid fuel cells. J Electrochem Soc 141 (10):2659–2668

Yang C, Costamagna P, Srinivasan S, Benziger J, Bocarsly A (2001) Approaches and technical challenges to high temperature operation of proton exchange membrane fuel cells. J Power Sources 103(1):1–9

Chapter 5
Catalysis in Production of Syngas, Hydrogen and Biofuels

5.1 Introduction

Efficient, economical, and environmental friendly chemical production usually needs the utilization of catalysis to run at rates that are acceptable. Most of the western standard of living is based on the ability of the chemical industry to convert available raw materials effectively into fuels, chemicals and applicable energy. It is crucial to know the definition of catalysis. Catalyst is defined as a substance that takes part and speed up the rate of a chemical reaction without being consumed by the reaction. The science which is related to the rate of reactions is chemical kinetics, which also includes catalysis. However, reactions have to be thermodynamically favorable to occur where the reaction must have an ability to occur spontaneously and in other words their Gibbs-free energy must be negative. A catalyst does not change the position of chemical equilibrium, but it speeds up the tendency of the reaction towards equilibrium.

Thermodynamics characterize the thing that will be possible, and catalysis together with kinetics illustrate how fast it is possible. Industrial catalysis, particularly heterogeneous catalysis, have developed widely simultaneously with the modern methods of industrial petroleum refining and has relentlessly permitted more complex utilization of feedstocks for a growing range of and request for petroleum products (Thomas et al. 2009; Nørskov et al. 2009). Similarly, an extensive part of the current diesel and petrol is prepared through respectively catalytic hydrocracking and fluid catalytic cracking of heavier fractions from the crude distillation; the hydrodesulphurisation of fuels to make sulphur-free fuels; or the preparation of hydrogen through steam reforming of natural gas (Strunk et al. 2009). The shift towards using of biomass as a feedstock for fuels and chemicals demands the development of new catalytic chemistry (Nørskov et al. 2009). Petroleum about completely insufficient of chemical functionalities that could be attacked without difficulty. This is good for fuels that use for internal combustion

© Springer International Publishing AG 2017
S. Bagheri, *Catalysis for Green Energy and Technology*,
Green Energy and Technology, DOI 10.1007/978-3-319-43104-8_5

engines. In addition to that, most refinery processes are run at high temperatures and with particular catalysts to activate the petroleum compounds.

Nevertheless, biomass is from a petrochemical perspective over-functionalized especially with respect to oxygen, and thus the conversion into appropriate fuels or chemicals commonly needs defunctionalization of most or all of these functionalities, rely upon their feedstock type. Catalysis can be an even higher concern as a discipline, and a thoughtful knowledge of the science behind it, additionally controlling it on a technical scale will be of extremely importance for the future chemical industry.

5.2   Catalysis in Production of Syngas

The word Syngas which refers to synthesis gas is used for illustrating a mixture of fuel gas with the composition of H_2 and CO, as well as small quantity of CO_2 and CH_4. The term "syngas" originates through its utilization as an intermediate in producing synthetic natural gas (SNG) (Beychok 1975) and also creating ammonia or methanol. The major usage of syngas is in the application of electricity production. Syngas is commonly generated from gasification, combustible and frequently utilized as fuel for internal combustion engines (Chandolias 2014; Boehman and Corre 2008). Syngas has not more than half the energy density of natural gas.

Syngas might be formed from divers of sources, inclusive of coal, biomass, natural gas, or any other feedstock made up of hydrocarbons, through reaction using steam via steam reforming, carbon dioxide via dry reforming and also oxygen via partial oxidation. The selection of one specific raw material relies upon the price and abundance of the feedstock, and use of syngas on downstream. Syngas is a very important intermediate which was used in the production of hydrogen, methanol, ammonia, and any other synthetic hydrocarbon fuels. In addition, it acts as intermediate in the production of synthetic petroleum for fuel or lubricant purposes through the Fischer–Tropsch (FT) process and formerly in the production of gasoline process from Mobil methanol. The strategy for the production purpose involves steam reforming of either NG or liquid hydrocarbons for the production of hydrogen, and also the gasification of either coal or biomass. Because of syngas was used as an intermediate to be used for numbers of different routes and aim, its structure and especially the ratios of H_2/CO are also very distinct (Moulijn et al. 2001).

For instance, the composition of both CO_2 and CO that act as the reactants in the reaction to synthesize methanol is defined by a module, $M = (H_2 - CO_2)/(CO + CO_2)$, and it is supposed to be near 2.0 mol/mole. Rather than for FT synthesis Gas to Liquid (GTL) applications, where the CO_2 is not one of the reactant, the ratio of the H_2/CO as syngas compositions should be about 2.0 mol/mole. Meanwhile, for the production of aldehydes through olefins hydroformylation, the optimal ratio of H_2/CO must be reduced to 1.0 mol/mole.

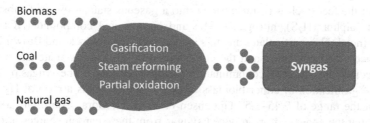

Fig. 5.1 Syngas production routes

Frequently, for the purpose of carbonylation process, the CO used must be pure. Nevertheless, the oil refining treatments and the synthesis of ammonia are the two major applications of syngas, which need the maximization on the production of H_2 (Fig. 5.1).

The production of syngas (H_2/CO) can be carried out by any appropriate source of carbon, yet the main typical raw materials or feedstocks are coal and NG. If biomass is used as the source of carbon, a greenhouse gas-neutral fuels and also chemicals might be created. For countries that have abundance of carbon sources (natural gas, coal, or biomass), the syngas production processes will then have the capability to minimize their dependence on foreign oil, maximize energy security and provide employment. The design part of the production of syngas is therefore very important for the profitability of the whole production plant. Nevertheless, the design of the production plant of syngas will mostly rely upon the accessible raw materials or feedstock and also the downstream utilization of the syngas. The excellent combination of all the production processes and utilization of energy is important for the effectiveness and economy of the plant (De Klerk and Refining 2011). Surely, the raw materials or feedstock abundance and cost has additionally been essential for size of plant, design of plant, and overall economy.

The technology for production of syngas is commonly divided into two categories, which are reforming and gasification. The utilization of reforming strategy takes place when any light liquid or gaseous feedstocks were subjected to be converted into syngas. Whilst, the strategy using gasification is utilized when any heavy liquid or solid feedstocks wants to be converted to syngas. When gasification taken place, the source of carbon is mixed with either oxygen or steam to produce gases majorly composed of H_2, CO, CO_2 and methane. The composition of these gases relies upon few parameters, for instance, the moisture and composition of the utilized feedstock, the medium of gasification: either air, steam or oxygen and the conditions of the reaction in terms of their temperature and pressure. Additionally, utilization of the gasifier and technology of the gasification reaction also affects the composition of the generated gases (Corella et al. 2008). Generally, there are three types of gasifier, which are fixed bed (bubbling or circulating), fluidized bed (downdraft and updraft) and entrained flow gasifiers, in which they have their own advantages and also disadvantages (Bridgwater 1995).

When the feedstock was transformed into a gaseous state, unwanted substances such as sulphur (H_2S), nitrogen (NH_3) and halogenated compound (HCl), also volatile metal (Na), particulates and tars are taken out (Woolcock and Brown 2013). The process of gasification and the composition of the starting materials or feedstock decides the degree of contamination (Bridgwater 1995). The syngas produced from the gasification of either biomass or coal commonly has a ratio of H_2/CO in between the range of 0.45–1.5 (Tijmensen et al. 2002; Göransson et al. 2011). It proved that the generated or produced syngas from these carbon sources are much richer in CO and poorer in H_2 than generated syngas from natural gas.

The commonly used strategies in the production of syngas are:

   i. Steam reforming (SR)
  ii. Partial oxidation (POx)
 iii. Auto thermal reforming (ATR).

5.2.1 Steam Reforming (SR)

Steam reforming is a catalytic and energy effective technology for forming a syngas that is richer in H_2 from light hydrocarbon feedstocks such as NG, LPG or naphtha. The prevailing commercial technology for production of syngas is steam methane reforming (SMR) from NG, where the process will catalytically mixed methane and steam, and then they will be endothermically transformed to H_2 and CO. SR is a technology that is commonly used for generating syngas. Reaction between light desulphurized hydrocarbons and steam; example, the steam reforming process of methane is expressed in Eq. 5.1:

$$CH_4 + H_2O = CO + 3H_2 \quad \Delta H°298\,K = +206\,kJ/mol \qquad (5.1)$$

When used for production of H_2, the steam reforming step is proceeded by a Water Gas Shift (WGS) step for the CO conversion as expressed in Eq. 5.2:

$$CO + H_2O = CO_2 + H_2 \quad \Delta H°298\,K = +41\,kJ/mol \qquad (5.2)$$

SR units consist of two parts, that is a radiant and also a convective area. The reforming processes occur within the radiated area. Within the convective area, heat is retrieved from the hot product gases in order to preheat the feedstocks reactants and producing superheated steam.

The reforming reaction using steam is endothermic reaction and they was catalyzed by Ni based materials where the deposited Ni species on the supports of ceramic that was made up of alumina or magnesium spinels (Rostrup-Nielsen 1993). Some reviews on the potential usage of noble metal based catalyst also including alkaline earth species are discussed in literature (Ghenciu 2002). The catalysts are contained within the tubes was then added into the radiant furnace. The diameter of the catalytic tubes is in the range between 3 and 5 in. and also in the range of 6–13 m length.

Usually, the temperatures of gas at the end of the tubes are more than 800 °C and their pressures are made up in the range of 15–30 barg. One necessity of reforming process using steam is described by the thermo-mechanical resistance of the tubes, whose skin temperature maintains at the values in the range between 100–150 °C more than those of the reaction environment for allowing high heat transfer rates. Due to that reason, the tubes are covered using alloys with a high content of Cr and Ni which is around 25–35% and their positioning in the radiant furnace is decided by both the needs of increasing heat fluxes towards the reaction zone and the necessity on the avoidance of impingement between tubes and flames formed by the burners. This impingement might result in rapid failure of the tubes (Ahmed and Krumpelt 2001). The feedstock made up of hydrocarbons is introduced into the tubes subsequently after reacted with steam at the ratio of steam/carbon is higher than 2.3 volume/volume, more often higher than 2.7 volume/volume. In order for the hydrocarbon reactions to stop and also to prevent the reaction of carbon formation, the excess steam is needed as described in Eqs. (5.3)–(5.5).

$$C_nH_{2n+2} = nC + (n+1)H_2 \qquad (5.3)$$

$$2CO = C + CO_2 \qquad (5.4)$$

$$CO + H_2 = C + H_2O \qquad (5.5)$$

Carbon and soot production reactions might cause the increment in pressure drop, deactivation of catalyst and reduction in the rate of reaction, results in crucial heat transfer difficulty and tube problems. Higher hydrocarbons are highly reactive towards reaction 5.3 when compared to methane which is one of light hydrocarbon. Because of that reason, once in a while, these C_{2+} molecules are converted inside an adiabatic pre-reformer unit (Rostrup-Nielsen et al. 1998). This unit could be arranged and conducted at relatively low temperatures (ca. 550 °C) results in numbers of advantages such as reduction in the size of the reformer furnace or increment in the production capability (Joensen and Rostrup-Nielsen 2002).

5.2.2 Partial Oxidation (POx)

The other technology for production of syngas is a partial oxidation (POx), that will exothermically convert methane and oxygen to syngas. These two technologies (SR and POx) constitutionally generate syngas with a hugely different ratio of H_2/CO which are about 3–5 in SMR and about 1.6–1.9 in POx (Rostrup-Nielsen 1994; Speight 2010). POx can be carry out by either catalytically and non-catalytically.

5.2.2.1 The Non-catalytic Partial Oxidation (POx)

Non-catalytic POx is a method used with a unique chance of using heavy hydrocarbon as the feedstock and generates a syngas with richer CO at temperatures in the range of 1100–1400 °C. Its energy effectiveness is lower than steam reforming. The chemistry behind the POx technology is based on the partial combustion of fuels that in the case of methane is expressed in Eq. 5.6:

$$CH_4 + 1/2\,O_2 = CO + 2H_2 \tag{5.6}$$

Nevertheless, non-catalytic POx is usually used for generating syngas from heavy hydrocarbons, inclusive of deasphalter pitch and petroleum coke. These are pre-heated and then react with oxygen in a burner. An outflow which consists of several numbers of soot that relies on the composition of the feedstocks will be generated in a combustion chamber that have a very high temperature right after ignition process. Reactor exit gas temperatures are usually enclosed between 1200–1400 °C. The generated syngas has to be cooled and cleaned in a "washing" section to get rid of the impurities.

The high temperature, which is in the range from 1400 to 1100 °C heat recovery in partial oxidation process is not very effective. When compared to the steam reforming process, the advantage POx have been their ability of using a "low value" feedstock, even if the feedstock consists of sulphur and other compounds that would harm the SR catalysts. Now, the primary usages of POx are: (i) in production of H_2 for refinery applications, (ii) production of syngas from coal and (iii) in production of electric energy from petroleum coke and deasphalter bottoms, via large Integrated Gas Turbine Combined Cycles (IGCC).

5.2.2.2 The Short Contact Time—Catalytic Partial Oxidation (SCT-CPO)

Early perception on the improvement of short contact time hydrocarbon oxidation procedures were depicted in the years 1992–1993 (Byrd and Hickman 1992; Leeflang et al. 1992). These procedures have been strongly examined from that point forward, and the scientific articles that have been published each year on this topic, is still high. They are created by striking for a couple of milliseconds, gaseous premixed reactant streams with greatly hot catalytic surfaces. The quick and specific chemistry that is begun is limited inside a thin (<1 mm) solid–gas inter-phase zone encompassing the catalyst particles. Here, the molecules spend 10–6 s at temperatures variable between 600–1200 °C.

One main problem of the technological exploitation is in the chances of preventing the propagation of reactions into the gas stage, that needs to stay at a "moderately low" temperature. This condition supports the arrangement of main reaction products which are CO and H_2 hindering chain reactions. For sure, some experimental reviews whose outcomes have been somewhat depicted in literature

(Schwiedernoch et al. 2003; Oberhammer et al. 1993; Basini et al. 2000; Grunwaldt et al. 2001; Jakowski et al. 2002; Bizzi et al. 2004) show that fractional and total oxidation products are specifically created through parallel and contending surface reactions and that the arrangement of partial oxidation products is supported under SCT conditions because of the high surface temperatures. By appropriate decision of the working conditions, surface temperatures are locally considerably higher than those anticipated by thermodynamic equilibrium calculations accepting adiabatic reactors.

The event of the reactions in these local environments decides in some cases conversion and selectivity values higher than those anticipated by the thermodynamic equilibrium at the reactor exit temperatures (Grunwaldt et al. 2001). Also, the high surface temperatures restrain catalyst deactivation phenomenon corresponds to chemical poison effects (Grasselli et al. 2005). For these and other related reasons, this chemical process is done in very small reactors having a very high adaptability towards reactant flow variations. It has additionally been found that few hydrocarbon feedstocks, even containing sulphur and aromatic compounds can be fed to a SCT-CPO reactor for creating synthesis gas. The Short Contact Time—Catalytic partial Oxidation technology, among the different proposed solutions, has achieved enough dependability to advance its industrialization.

5.2.3 Auto Thermal Reforming (ATR)

Auto thermal reforming (ATR) is a third option, which can be viewed as a hybrid between the two past options in a single reactor. In the combustion zone, parts of the feed are combusted with oxygen, while in the reforming zone the rest of the feed and the generated CO_2 and H_2O are catalytically reformed to syngas. The energy needed for the endothermic reforming reactions is given out by the exothermic oxidation reactions from the combustion zone. ATR combines gaseous phase combustion reactions and catalytic steam/CO_2 reforming reactions; it is significantly less utilized than SR and POx but it is the ideal choice for integration with large scale MeOH production plants and GTL processes (Holladay et al. 2009).

In spite of the long term, R&D and industrialization effort that has prompted to the optimization of these technologies a relevant effort is still ongoing for defining new radical changes permitting a reduction of the capital and energy requirements of the syngas production step. ATR combines non-catalytic partial oxidation and catalytic steam and CO_2 reforming of light and highly de-sulphurated NG in a single reactor. The process was created in the late 1950s by Haldor Topsøe A/S, predominantly to produce syngas for methanol and ammonia plants and furthermore for the Fischer-Tropsch synthesis (Christensen and Primdahl 1994; Aasberg-Petersen et al. 2001). The NG is blended at high temperature with a mixture of oxygen and steam and ignited in a combustion chamber originating a sub-stoichiometric flame that can be expressed by Eq. 5.7.

$$CH_4 + 3/2\,O_2 = CO + 2H_2O \quad \Delta H°298\,K = -519\,kJ/mol \qquad (5.7)$$

Subsequently, steam and CO_2 reforming reactions 5.8 take place inside a catalytic bed situated underneath the combustion chamber.

$$CH_4 + CO_2 = 2CO + 2H_2 \quad \Delta H°298\,K = 247.0\,kJ/mol \qquad (5.8)$$

By proper modification of oxygen to carbon and steam to carbon ratios the partial combustion in the thermal zone supplies the heat for finishing the subsequent endothermic steam and CO_2 reforming reactions (Joensen and Rostrup-Nielsen 2002). The product gas composition at the exit of the reactor results very close to the thermodynamic equilibrium of an adiabatic reactor, particularly in huge scale processes (Rostrup-Nielsen 2000). ATR is additionally used as a "secondary reformer" (for bringing down the CH_4 residue) and it is put after a primary SR in syngas plants incorporated with ammonia synthesis reactors. In this situation, the "secondary" ATR is fed with the syngas created from SR and air.

5.3   Catalysis in Production of Hydrogen

Hydrogen is one of the main feedstocks utilised as a part of the chemical industry. It is an essential building block for the production of ammonia, and thus fertilizers, and of methanol, utilized as a part of the production of many polymers. Hydrogen is utilized as a part of the fabricate of two of the most essential chemical compounds made industrially, which are ammonia and methanol. It is likewise utilized as a part of the refining of oil, for instance, in reforming, one of the processes for getting high grade petrol and in removing sulphur compounds from petroleum which would somehow damage the catalytic converters fitted to cars.

In the future, hydrogen itself may get to be distinctly a standout amongst the most important fuels for cars as on burning it does not generate carbon dioxide, yet there are real issues to be overcome before it can be utilized as a part of along these lines, including its production, storage and distribution. The manufacture of hydrogen is the family of industrial techniques for producing hydrogen. Some of the techniques to be used for the generation of hydrogen are listed below.

Hydrogen production strategies:

- Reforming of hydrocarbons
- Reforming of biomass
- Electrolysis
- Thermolysis
- Photolytic transformation
- Biological transformation

At present, the dominant technology for direct manufacturing is steam reforming from hydrocarbons. Starting at 1999, the large number of hydrogen ($\sim 95\%$) is created from fossil fuels by steam reforming or partial oxidation of methane and coal gasification with just a little amount by different routes, for example, biomass gasification or electrolysis of water. By far the most essential process for making hydrogen is by *steam reforming*. Hydrogen is comprehensively used in the deoxygenation of biomass inferred feedstocks; nonetheless, it is desirable to limit the degree that external H_2 is needed for production of biofuels, particularly if this H_2 is generated from petroleum based feedstocks. The early stage of transforming into fuels from biomass can continue in all cases without the contribution of hydrogen. The management of hydrogen turns into a more basic issue in the next stage of biofuel generation wherein the upgradeable platforms are changed to liquid hydrocarbon fuels.

Usually, two catalytic alternatives are possible to be utilized for the generation of renewable hydrogen from biomass. The most essential choice is through the gasification stage, where they can be mixed with the water-gas shift reaction to support hydrogen generation from CO and H_2O (Eq. 5.9). The second alternative is aqueous phase reforming of oxygenates, for example, polyols 62–64 (Eqs. 5.10 and 5.11). These systems can be considered analogous to the generation of hydrogen from petroleum feedstocks using steam reforming alternatives.

$$CO + H_2O \rightarrow CO_2 + H_2 \tag{5.9}$$

$$C_3H_8O_3 + 3H_2O \rightarrow 3CO_2 + 7H_2 \tag{5.10}$$

$$C_6H_{14}O_6 + 6H_2O \rightarrow 6CO_2 + 13H_2 \tag{5.11}$$

The main parts of the process are the transformation of a material containing carbon to a blend of CO and H_2 took after by the change of CO to CO_2. Currently, methane or any other hydrocarbons which are light that generated from NG or oil were utilized. The gas or vapor is blended with an expansive abundance of steam and went via funnels consisting of nickel oxide (which is reduced to nickel amid the reaction), upheld on alumina, in a furnace which works at high temperatures:

$$CH_4 + H_2O \rightarrow 3H_2 + CO \quad \Delta H° = +210 \, kJ \, mol^{-1} \tag{5.12}$$

The reaction is endothermic and joined by an expansion in volume. It is along these lines supported by high temperatures and by low partial pressures. The reaction is likewise supported by a high ratio of steam: hydrocarbon. This expands the yield; however, the operating cost are increased. The high ratio additionally lessens the measure of carbon deposited that will diminish the productivity of the catalyst. The best approach to diminish carbon deposition was observed to be impregnation of the catalyst with potassium carbonate. In the next stage of the process, carbon monoxide is changed over to carbon dioxide:

$$CO + H_2O \rightarrow H_2 + CO_2 \quad \Delta H° = -42\,kJ\,mol^{-1} \qquad (5.13)$$

There is a critical change in the selection of fuel utilized in the reformer. Rather than a hydrocarbon gas, in many countries such as China, coal is more accessible. Different hydrocarbon gases (commonly used methane) that were oil-based are hard to find and should be imported.

Research is being attempted to see whether biomass can replace oil or coal to be utilized efficiently for hydrogen production. The problem is that the energy utilized as a part of gathering it, transporting it to the place of utilization can be high with respect to the savings in changing to biomass. Hydrogen is possible an environmentally attractive fuel for the future. When it was consumed to create energy, the product formed is only water. The route to create the gas is in the reverse process, the electrolysis of water. The general overall equation is:

$$2H_2O \rightarrow 2H_2 + O_2 \qquad (5.14)$$

Nevertheless, this process requires supply of electricity from power stations. If the station utilizes fossil fuels, then the objective specifically to prevent generation of carbon dioxide from fuel is defeated. Different types of creating power, such as wind and geothermal, do not have this drawback. Much research is being attempted to create fuel cells where, about portions of the energy from the reaction between hydrogen and oxygen to deliver water is discharged as an electrical potential. One such fuel cell is the PEM cell, where PEM stands for Polymer Electrolyte Membrane, which portray the activity of the cell. A flood of hydrogen is conveyed to the anode on one side of a membrane (a fluorinated polymer based on poly (tetrafluoroethene) (PTFE) or carbon fibre.

$$H_2 \rightarrow 2H^+ + 2e^- \qquad (5.15)$$

The protons saturate the membrane and then will react with oxygen at the cathode:

$$4H^+ + O_2 + 4e^- \rightarrow 2H_2O \qquad (5.16)$$

Both reactions are catalyzed by platinum, which is as nanoparticles installed in the electrodes. Extremely pure hydrogen must be used in the reaction, because the catalyst will be harmed quickly, even if there is only a little amount of CO from its fabricate. Numbers of research is are being directed to discover methods for enhancing the lifetime of the catalysts utilized. In spite of the fact that the fuel cells are being experimented in a few cars and other different vehicles, there are pragmatic troubles in the appropriation and capacity of hydrogen. One approach is changing over a liquid fuel into hydrogen, in situ in the car. For instance, methanol is being utilized as a part of experimental cars. The vapor is changed into H_2 and CO_2 by a reforming reaction, same like the process portrayed above for the large-scale production of hydrogen. This needs an extremely high level of engineering skills to generate conversion units which are sufficiently light for a car yet sufficiently strong to withstand every one of the issues brought about by continuous vibrations.

Research is in progress to transform natural gas to hydrogen using mini-reformers. A huge measure of research is likewise being attempted on utilizing sunlight as the energy source, through bio photolysis. This includes the manufacturing of algae in water via photosynthesis, trailed by bacterial deterioration of the algae to generate hydrogen. A critical disclosure is by denying the algae of sulphur, typical photosynthesis is restrained and rather an enzyme is initiated and hydrogen, not oxygen is generated in light. Current research is concerned in making these processes more effective.

A catalyst brings down the number of energy required to part those molecules, and platinum is super great at this. Be that as it may, platinum is likewise ridiculously expensive—much excessively expensive for widespread utilize in the production of hydrogen. With the help of platinum catalysts, it is conceivable to proficiently generate hydrogen. Nevertheless, this metal is uncommon and really expensive. Researchers have found an option that is similarly as great, yet less expensive.

The mineral pentlandite is a potential new catalyst for the production of hydrogen. As depicted in the journal *Nature Communications*, it works similarly as effective as the platinum electrodes generally utilize today. As opposed to platinum, pentlandite is affordable and discovered as often as possible on Earth. Notwithstanding platinum, there are various different substances that can catalyze the reaction of water to hydrogen and oxygen and do not contain any valuable metals. Among such compounds are the supposed metal chalcogenides. For the most part, nevertheless, these non-metallic materials are unmistakably poor conductors of electrons and are thus ineffective catalysts.

Pentlandite comprises of iron, nickel, and sulphur. Its structure is like the active centres of hydrogenases, which are hydrogen-producing enzymes, as found, for instance, in green algae. In the present research, the scientists compared the rate of the production of naturally acquired and artificially delivered pentlandite with platinum and other non-metallic catalysts. Artificial pentlandite and platinum turn out to be similarly great catalysts, with a performance that outperforms that of the various materials tried. The mineral synthesized in the lab generated hydrogen a great deal more effectively than the naturally discovered variant. The reason: Inclusions of magnesium and silicon in natural pentlandite diminish its conductivity. The researchers called the yield of artificial pentlandite "shockingly high," and the rate of synthesis also stayed stable for quite a while.

The mineral has another advantage contrasted with other non-valuable-metal materials. It has a more prominent, active surface area to which the reacting substances can dock. In other non-valuable-metal materials, this surface must be made utilizing complex strategies by applying the catalyst to an electrode as nanoparticles.

5.4   Catalysis in Production of Biofuels

The utilization of renewable sources to generate biofuels use for transportation purpose is an essential mitigation strategy for the effect of fossil fuels on the environment. However, there is an issue arise from that strategy regarding their costs. The cost to generate biofuel is very high, and without any subsidies given to this industry, it is quite impossible to be carried out. Biofuels can be generated by various processes, comprehensively divided to biochemical and chemical. Biological catalysts called as enzymes were used in biochemical processes and chemical catalysts were used in the chemical processes, both for the same purpose which is to convert crude materials into fuels.

New technologies are designed and tried in research facilities on a regular basis. Nevertheless, to make them industry prepared requires advanced work and still the process/chemical/enzyme may not be proper in light of cost. Some of the recent advances in these technologies that can possibly make the production of biofuel more economical.

5.4.1   Biochemical Processes

Biodiesel

Phospholipase and also lipase are two of the enzymes that plays an important role for biodiesel. Biodiesel that was generated from enzymatic processes have been commercializing a number of organizations. The major product that including biodiesel are fatty acid methyl esters which can be generated from the free fatty acids (FFA) and triacylglycerol by the help of lipase. The phospholipase is in charge of converting phospholipids to diacylglycerol, which turns into a substrate for the lipase. In the conventional processes, that involve the utilization of methanol and also catalysts, the quality of the biodiesel can only be enhanced if the free fatty acids and the phospholipids were evacuated before the reactions. The quantity of the product is high and the reaction spares chemical waste just because the enzyme lipase can use the phospholipids as substrates. One of the drawbacks on using enzyme is that their cost is too high and it limits their utilization on all feedstocks, especially clean plant oils.

One method for augmenting the life and subsequently bringing down the cost of the enzymes is by doing the immobilization of enzyme on solid substrate to empower various cycles of utilization. The other method to solve the problem is to generate them in a more cost proficient system or to enhance their activity.

Cellulosic ethanol

Cellulases play an important role in generating of glucose from cellulose that will be used in aging to biofuels in the case of biomass conversion. In the course of recent years, exceptional investigations of issues encompassing the use of cellulosic feedstocks for the production of biofuel have been carried out. Despite the fact that in principle this process can use the huge amounts of biomass accessible from cultivating and devoted feedstocks, real issues of conversion are experienced. The cost of the enzymes required for deconstruction of the cellulose and hemicellulose into usable sugar streams is one of the real issues. A few methodologies were carried out by specialists, including modifying the structure of the cell walls to bring down the trouble of assimilation, discovering better enzymes, utilizing consolidated digestion and aging, and discovering better pre-treatment technologies to set up the feedstock for the enzymes.

Enzyme cost

Deconstruction of pre-treated biomass are carried out by blending of enzymes. The utilization of enzymes cost viably, the NREL has estimated that the cost of the enzymes ought to be $0.10 per gallon of biofuel. For as far back as 15 years, the extraordinary investigation on the production of enzyme platforms has obtained contagious mixtures of enzymes which do not meet these cost pre-requisites and in actually likewise need an enormous infrastructure for manufacture. A generally new technology uses genetically engineered plant seeds (mainly maize) to collect industrial enzymes. At scale, enzymes from this framework can cost lesser in order to generate and formulate as a result of low prerequisites for capital infrastructure. Despite the fact that the production of the plant seed system is more cost focused, it has not been tried at scale for viability. Other research efforts are in multifunctional enzymes and joined bioprocessing organisms, the last of which can break down plant polymers and in addition mature them into biofuels.

5.4.2   Chemical Catalysts

Chemical catalysis has been chosen as the suitable strategy for the proficient creation of transportation energizes from fossil carbon sources, thus it is very common for the strategy to be use in technology of biomass transformation. A considerable lot of the courses to the transformation of biomass include a few steps, including depolymerisation took after by separation and upgrading processes.

Pre-treatment and Deconstruction

Usually, the pre-treatment of biomass with the use of strong acid and base hydrolysis is carried out as the initial step and this step is a destructive process. Now, an ammonia based AFEX™ strategy was build up by researchers. The strategy involves the process of opens up the structure of carbohydrates that have been isolated from lignin by treatment with ammonia. The strategy enormously enhances the effectiveness of upgrading the enzymes. It is likely that other downstream catalytic methodologies will likewise be encouraged by pre-treatment.

Multifunctional Catalysts

Depolymerization processes will result in products that incorporate with materials that is not acceptable for the use as fuels such as light gases and highly oxygenated compounds. Therefore, isolation processes should take place for the materials upgrading purposes. In any case, the needed isolations processes and waste products managements from pre-treatment steps will increase the cost of biofuel generation and make the processes become more complex. These issues led the researchers to look for heterogeneous catalysts that can specifically change over biomass to liquid fuels. Consolidating biomass depolymerization and upgrading to liquids by deoxygenation in one step is especially alluring. Additionally, it is extra alluring with one step reforming process of light gases into liquid products. Recombination of intermediate products into high molecular weight compounds and tars can take place without being controlled by extra catalyst capacity. The immediate transformation needs a strategy for reaching the biomass with the catalyst either by the volatilization by pyrolysis or utilization of solvent.

Thus, it is very important to utilize the multifunctional catalyst frameworks that can mix acid and metals to overcome the issue mentioned earlier. It is hard to adjust the activity components of these frameworks to generate the ideal outcomes. The extra step of processes was needed to change the low value carbon oxides and acids, which elevated from high temperature gasification to become liquid fuels. The recombination reactions can be permitted by long contact times. The most widely recognized way to deal with this issue is to conduct reactions in various different zones without interstage separation. The most desirable catalysts are the catalysts that can advance conversion of feedstock even at low temperature.

Zeolite catalysts have revolutionized the processing of petroleum so it is not unpredictable, they have gotten a considerable measure of consideration as potential biomass catalysts. ZSM5 has ended up being the best of the commercial zeolites due to its selectivity to bring down molecular weight aromatic, deoxygenation activity, and low coking properties. As of late consideration has swung to the impacts of introducing cheaper base metals to the zeolite to enhance transformation and aromatic selectivity. Ni catalysts have gotten specific consideration. A current case is introducing Ni to ZSM5 to build up the yield of aromatic hydrocarbons while all the while expanding the transformation of oxygenates.

Research on treating the vapor phase from the pyrolysis of sawdust using the ZSM5 that was altered using Zr, Co and Fe as catalyst is being widely explored. The impacts of biomass pre-treatments were additionally investigated. It was found that the joined pre-treatments and utilization of a FeZSM5 extraordinarily enhanced aromatic yield contrasted to reaction with unmodified saw dust and ZSM5. Zeolite catalysts have a few disadvantages as far as restricted hydrothermal stability and a propensity for the smaller scale pores to plug with coke or different deposits.

5.5   Application of Fischer–Tropsch Synthesis in Biomass to Liquid Conversion

The Fischer–Tropsch process is an accumulation of chemical reactions that changes over a mixture of carbon monoxide and hydrogen into liquid hydrocarbons. The Fischer-Tropsch-synthesis (FTS) was at first introduced by German researchers Franz Fischer and Hans Tropsch in the 1920s. By steam gasification of coal, steam reforming of natural gas or gasification of biomass, a synthesis gas (syn-gas) comprising of CO and H_2 can be achieved after some work-up, which is utilized for the FTS yielding principally straight-chained alkanes. At the season of Fischer and Tropsch coal was the applicable raw material for generating syn-gas, giving a method for transforming cheap coal into liquid fuels—so-called coal-to-liquids (CTL). This has been utilized by Germany amid World War II, and South African company Sasol has been utilizing it for more than 50 years—the plant in Sasolburg is the greatest point CO_2-emitter in the world (Kintisch 2008).

An enormous body of literature is accessible in research on FTS catalysis, which is normally performed by metal-oxide-supported metals at 150–330 bar. While supported ruthenium has been appeared to be the most active metal for FTS (Simonetti et al. 2007) iron and cobalt are typically utilized industrially, since the platinum group metals are generally excessively costly, making it impossible to take into account for commercial use at this scale. The Co or Fe may then be doped with smaller amounts of other metals as promoters, for example, Mn, Ni, Pt, Ru, K, or Ce. Aside from water, straight-chain alkanes are obtained from FTS because of the reaction mechanism of chain propagation, which is perfect for diesel fuels in the events that they have the correct carbon number. Another essential element of the reaction is that for a specific carbon chain length i, the weight fraction in the product mixture is given from the statistical probability of chain growth, α, by:

$$W_i = i \cdot (1 - \alpha)^2 \cdot \alpha^{i-1} \tag{5.17}$$

This condition is known as the Anderson-Schultz-Flory-distribution (ASF) (Davis 2007; Kaneko et al. 2005). The product distribution as a function of the chain growth probability, α, is appeared for specific groups of alkanes. Frequently, the selectivity is described, for example, as the yield of hydrocarbons with a chain-length of no less than 5 carbon atoms, C_{5+}. This is the group of hydrocarbons, most applicable regarding of refinery processing and fuel production. FTS-catalysts always form methane to some degree, and a low selectivity to methane and also the light hydrocarbon gases are typically required (scientifically this implies a higher estimation of α in the ASF) (Davis 2007; Kaneko et al. 2005). It takes after mathematically from the ASF that methane is dependably the product generated in the most outstanding amount on a molar basis; however the weight fraction is small because of the large molar weight of the heavier paraffins formed at high values of α (Davis 2007).

The essential catalysts for FTS Co and Fe generally have $\alpha \approx 0.7$–0.9, which can be advanced by addition of promoters to the primary metal. The biggest part of a diesel pool of C_{12}-C_{18}-hydrocarbons is accomplished with $\alpha \approx 0.88$. Iron has been appeared to be the more active metal under the most serious conditions, i.e. at higher temperature, pressure and space velocity, while oppositely cobalt is the more active catalyst in less compelling conditions (Davis 2007; Patzlaff et al. 1999). Notwithstanding straight-chain hydrocarbons, impurities of aldehydes, alcohols, and fatty acids are also generated. Such oxygenated by-products ought to later be decreased by HDO if utilized for diesel synthesis. Light isomerization might be important to accomplish adequately good cold properties (Gamba et al. 2010).

The FTS catalysts are delicate to harms—particularly sulphur, which generate sulphides with the catalyst metal if present over ppm-level. The particle-sizes and structure of the catalyst metal, and additionally the microscopic condition in which they are utilized, influences the general reaction behavior. By supporting nanoparticles of Co, Fe and their alloys inside carbon nanotubes demonstrated that activity was upgraded, and the C_{5+}-selectivity could be improved also, which was credited to a more drawn out contact time with the catalysts particles in the bound condition of the nanotubes. Co had the most astounding C_{5+}-selectivity, and Co particles up to the span of the carbon nanotube diameter of 12 nm could be specifically situated inside the nanotubes. Lessened sintering (irreversible deactivation) inside the nanotubes because of the constrainment and stronger support interaction upgraded the FTS catalyst lifetime.

Larger Co particles had higher C_{5+}-selectivity yet bring down general activity because of a lower surface area of larger particles (Tavasoli et al. 2010, 2009). The ideal particle size is around 6–8 nm for Co as the C_{5+}-selectivity does not increase over this particle size (Bezemer et al. 2006). Oxidation of the Co crystallites happens amid reaction and impedes activity, but that as it may, the oxidized Co might be diminished to recover the activity (Tavasoli et al. 2010). The working Co catalyst have been appeared to be metallic (Bezemer et al. 2006). The Co-Fe alloys had high selectivity towards C_{2+}-alcohols as up to 26% of the carbon could be changed over to alcohol (Tavasoli et al. 2009).

Constrainment of active sites inside micro- or mesoporous materials, for example zeolites, can drastically modify the behavior of the FTS catalysts and in this way change the product distribution far from ASF and towards a more particular range of products, for example because of shape and size selectivity (Chen et al. 2008). Support acidity and micro porosity might be utilized to specifically isomerise products too (Liu et al. 2009). The pressure has an impact on the value of α as higher pressures commonly resulting higher chain-growth probabilities (Davis 2007). Advancement by alloying with transition metals and doping with alkalis have likewise been examined. For instance, little measures of K in a Co catalyst expanded the diminishment temperature and diminished the activity, however upgraded C_{5+}-selectivity and the selectivity towards olefins rather than paraffins.

Potassium may, in any case, additionally improve the activity, which is reliant on temperature (Davis 2007). Ru diminished the reduction temperature on Co and expanded activity and C_{5+}-selectivity (Trépanier et al. 2009). Mo addition to Fe decreased the deactivation by sintering of the active iron particles and improved the development of iron carbides (Ma et al. 2006)—some portion of the mechanism for FTS over Fe includes arrangement of carbides (Herranz et al. 2006; Blanchard et al. 2010). Ce and Mn promotors upgrade the formation of active carbide species for the chain development over the Fe catalyst, while Ce is likewise a structural promotor for Fe (Herranz et al. 2006). Iron catalysts, contingent upon advancement, can likewise catalyze the water-gas-shift equilibrium amid FTS (Herranz et al. 2006). A few researchers have attempted to enhance the generation of alcohols, aldehyde, or ketones (Tavasoli et al. 2009; Christensen et al. 2009; Durham et al. 2010). The generation of longer-chain alcohols or aldehydes might be another imminent advancement inside FTS, since overhauling and drop-in blend for either petrol, jet fuel or diesel can now be custom-made, or even be guided towards the generation of bulk chemicals via FTS. As was described by Mentzel and co-authors, higher alcohols may even be a better feedstock for generating petrol range fuel from the methanol-to-hydrocarbons reaction over acidic zeolites (Mentzel et al. 2009).

References

Aasberg-Petersen K, Hansen J-HB, Christensen T, Dybkjaer I, Christensen PS, Nielsen CS, Madsen SW, Rostrup-Nielsen J (2001) Technologies for large-scale gas conversion. Appl Catal A 221(1):379–387

Ahmed S, Krumpelt M (2001) Hydrogen from hydrocarbon fuels for fuel cells. Int J Hydrogen Energy 26(4):291–301

Basini L, Guarinoni A, Aragno A (2000) Molecular and temperature aspects in catalytic partial oxidation of methane. J Catal 190(2):284–295

Beychok MR (1975) Process and environmental technology for producing SNG and liquid fuels

Bezemer GL, Bitter JH, Kuipers HP, Oosterbeek H, Holewijn JE, Xu X, Kapteijn F, van Dillen AJ, de Jong KP (2006) Cobalt particle size effects in the Fischer–Tropsch reaction studied with carbon nanofiber supported catalysts. J Am Chem Soc 128(12):3956–3964

Bizzi M, Saracco G, Schwiedernoch R, Deutschmann O (2004) Modeling the partial oxidation of methane in a fixed bed with detailed chemistry. AIChE J 50(6):1289–1299

Blanchard J, Abatzoglou N, Eslahpazir-Esfandabadi R, Gitzhofer F (2010) Fischer–Tropsch synthesis in a slurry reactor using a nanoiron carbide catalyst produced by a plasma spray technique. Ind Eng Chem Res 49(15):6948–6955

Boehman AL, Corre OL (2008) Combustion of syngas in internal combustion engines. Combust Sci Technol 180(6):1193–1206

Bridgwater A (1995) The technical and economic feasibility of biomass gasification for power generation. Fuel 74(5):631–653

Byrd JW, Hickman KA (1992) Do outside directors monitor managers? Evidence from tender offer bids. J Financ Econ 32(2):195–221

Chandolias K (2014) Rapid bio-methanation of syngas by high cell-density in reverse membrane bioreactors

Chen W, Fan Z, Pan X, Bao X (2008) Effect of confinement in carbon nanotubes on the activity of Fischer–Tropsch iron catalyst. J Am Chem Soc 130(29):9414–9419

Christensen JM, Mortensen PM, Trane R, Jensen PA, Jensen AD (2009) Effects of H$_2$S and process conditions in the synthesis of mixed alcohols from syngas over alkali promoted cobalt-molybdenum sulfide. Appl Catal A 366(1):29–43

Christensen T, Primdahl I (1994) Improve syngas production using autothermal reforming. Hydrocarbon Process (United States) 73(3)

Corella J, Toledo J-M, Molina G (2008) Biomass gasification with pure steam in fluidised bed: 12 variables that affect the effectiveness of the biomass gasifier. Int J Oil Gas Coal Technol 1(1–2):194–207

Davis BH (2007) Fischer–Tropsch synthesis: comparison of performances of iron and cobalt catalysts. Ind Eng Chem Res 46(26):8938–8945

De Klerk A, Refining F-T (2011) Weinheim. Wiley-VCH Verlag GmbH & Co. KGaA, Germany

Durham E, Zhang S, Roberts C (2010) Diesel-length aldehydes and ketones via supercritical Fischer Tropsch synthesis on an iron catalyst. Appl Catal A 386(1):65–73

Gamba S, Pellegrini LA, Calemma V, Gambaro C (2010) Liquid fuels from Fischer–Tropsch wax hydrocracking: Isomer distribution. Catal Today 156(1):58–64

Ghenciu AF (2002) Review of fuel processing catalysts for hydrogen production in PEM fuel cell systems. Curr Opin Solid State Mater Sci 6(5):389–399

Göransson K, Söderlind U, He J, Zhang W (2011) Review of syngas production via biomass DFBGs. Renew Sustain Energy Rev 15(1):482–492

Grasselli F, Basini G, Bussolati S, Bianco F (2005) Cobalt chloride, a hypoxia-mimicking agent, modulates redox status and functional parameters of cultured swine granulosa cells. Reprod Fertil Dev 17(7):715–720

Grunwaldt J-D, Basini L, Clausen BS (2001) In situ EXAFS study of Rh/Al2O3 catalysts for catalytic partial oxidation of methane. J Catal 200(2):321–329

Herranz T, Rojas S, Pérez-Alonso FJ, Ojeda M, Terreros P, Fierro JLG (2006) Genesis of iron carbides and their role in the synthesis of hydrocarbons from synthesis gas. J Catal 243(1): 199–211

Holladay JD, Hu J, King DL, Wang Y (2009) An overview of hydrogen production technologies. Catal Today 139(4):244–260

Jakowski N, Wehrenpfennig A, Heise S, Reigber C, Lühr H, Grunwaldt L, Meehan T (2002) GPS radio occultation measurements of the ionosphere from CHAMP: early results. Geophys Res Lett 29(10)

Joensen F, Rostrup-Nielsen JR (2002) Conversion of hydrocarbons and alcohols for fuel cells. J Power Sources 105(2):195–201

Kaneko T, Derbyshire F, Makino E, Gray D, Tamura M, Li K (2005) Coal liquefaction. Ullmann's Encycl Ind Chem

Kintisch E (2008) The greening of synfuels. Science 320(5874):306–308

Leeflang EP, Liu W-M, Hashimoto C, Choudary PV, Schmid CW (1992) Phylogenetic evidence for multiple Alu source genes. J Mol Evol 35(1):7–16

Liu S, Gujar AC, Thomas P, Toghiani H, White MG (2009) Synthesis of gasoline-range hydrocarbons over Mo/HZSM-5 catalysts. Appl Catal A 357(1):18–25

Ma W, Kugler EL, Wright J, Dadyburjor DB (2006) Mo–Fe catalysts supported on activated carbon for synthesis of liquid fuels by the Fischer–Tropsch process: effect of Mo addition on reducibility, activity, and hydrocarbon selectivity. Energy Fuels 20(6):2299–2307

Mentzel UV, Shunmugavel S, Hruby SL, Christensen CH, Holm MS (2009) High yield of liquid range olefins obtained by converting i-propanol over zeolite H-ZSM-5. J Am Chem Soc 131 (46):17009–17013

Moulijn JA, Van Diepen A, Kapteijn F (2001) Catalyst deactivation: is it predictable? What to do? Appl Catal A 212(1):3–16

Nørskov JK, Bligaard T, Rossmeisl J, Christensen CH (2009) Towards the computational design of solid catalysts. Nat Chem 1(1):37–46

Oberhammer F, Wilson J, Dive C, Morris I, Hickman J, Wakeling A, Walker PR, Sikorska M (1993) Apoptotic death in epithelial cells: cleavage of DNA to 300 and/or 50 kb fragments prior to or in the absence of internucleosomal fragmentation. EMBO J 12(9):3679

Patzlaff J, Liu Y, Graffmann C, Gaube J (1999) Studies on product distributions of iron and cobalt catalyzed Fischer–Tropsch synthesis. Appl Catal A 186(1):109–119

Rostrup-Nielsen JR (1993) Production of synthesis gas. Catal Today 18 (4):305–324

Rostrup-Nielsen JR (1994) Catalysis and large-scale conversion of natural gas. Catal Today 21(2–3):257–267

Rostrup-Nielsen JR (2000) New aspects of syngas production and use. Catal Today 63(2):159–164

Rostrup-Nielsen JR, Christensen TS, Dybkjaer I (1998) Steam reforming of liquid hydrocarbons. Stud Surf Sci Catal 113:81–95

Schwiedernoch R, Tischer S, Correa C, Deutschmann O (2003) Experimental and numerical study on the transient behavior of partial oxidation of methane in a catalytic Monolith. Chem Eng Sci 58(3):633–642

Simonetti DA, Rass-Hansen J, Kunkes EL, Soares RR, Dumesic JA (2007) Coupling of glycerol processing with Fischer–Tropsch synthesis for production of liquid fuels. Green Chem 9 (10):1073–1083

Speight JG (2010) The biofuels handbook, vol 5. Royal Society of Chemistry, UK

Strunk J, Kaehler K, Xia X, Comotti M, Schueth F, Reinecke T, Muhler M (2009) Au/ZnO as catalyst for methanol synthesis: the role of oxygen vacancies. Appl Catal A 359(1):121–128

Tavasoli A, Trépanier M, Abbaslou RMM, Dalai AK, Abatzoglou N (2009) Fischer-Tropsch synthesis on mono-and bimetallic Co and Fe catalysts supported on carbon nanotubes. Fuel Process Technol 90(12):1486–1494

Tavasoli A, Trépanier M, Dalai AK, Abatzoglou N (2010) Effects of confinement in carbon nanotubes on the activity, selectivity, and lifetime of Fischer–Tropsch Co/carbon nanotube catalysts. J Chem Eng Data 55(8):2757–2763

Thomas JM, Hernandez-Garrido JC, Bell RG (2009) A general strategy for the design of new solid catalysts for environmentally benign conversions. Top Catal 52(12):1630–1639

Tijmensen MJ, Faaij AP, Hamelinck CN, van Hardeveld MR (2002) Exploration of the possibilities for production of Fischer Tropsch liquids and power via biomass gasification. Biomass Bioenerg 23(2):129–152

Trépanier M, Tavasoli A, Dalai AK, Abatzoglou N (2009) Co, Ru and K loadings effects on the activity and selectivity of carbon nanotubes supported cobalt catalyst in Fischer-Tropsch synthesis. Appl Catal A 353(2):193–202

Woolcock PJ, Brown RC (2013) A review of cleaning technologies for biomass-derived syngas. Biomass Bioenerg 52:54–84

Chapter 6
Biodiesel and Green Diesel Production, Upgrading of Fats and Oils from Renewable Sources

6.1 Biodiesel Production

The alternative to conventional diesel or fossil diesel is biodiesel. It can be produced from either fats and oils (vegetable oils, used cooking oils). Production of biodiesel can be carried out chemically by esterification or transesterification reactions. These reactions take place between the fats or oil with short chain or low molecular weight alcohol such as ethanol and methanol. Commonly, ethanol is being used due to its low cost. However, when methanol is used in the reactions, greater conversion of fats and oils into biodiesel can be reached. Through transesterification reactions, the fats and oils will be converted into long chain mono alkyl esters which can be used as a fuel and being called biodiesel. The commonly called FAME or fatty acids methyl esters are referred to the converted long chain mono alkyl esters.

Roughly, 100 lb of biodiesel and 10 lb of glycerol will be formed when 100 lb of fats or oil were reacted with 10 lb of short chain alcohol in the presence of catalyst (NaOH or KOH). The transesterification reactions are commonly base catalysed even though it can be both, either bases or acids catalysed. When compared to acid catalyst, base catalyst provide lower reaction times and lower cost (Anastopoulos et al. 2009) (Fig. 6.1).

The major steps required to synthesise biodiesel are as follows:

(i) Feedstock pre-treatment

Virgin vegetable oils or recycled vegetable oils (yellow grease) are commonly used as the feedstock in production of biodiesel. Refinery of virgin oils was carried out, but not to a level of food grade. Removal of phospholipids and other plant matter is done by degumming, though refinement processes vary (Olivares et al. 2014). Meanwhile, the recycled oils are processed in order to remove impurities (dirt, water and charred food). Water can cause the triglycerides to hydrolyse during the

© Springer International Publishing AG 2017
S. Bagheri, *Catalysis for Green Energy and Technology*,
Green Energy and Technology, DOI 10.1007/978-3-319-43104-8_6

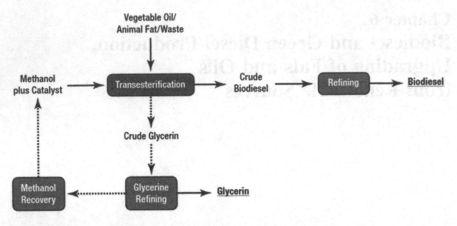

Fig. 6.1 Schematic of biodiesel production path

reactions, giving salts of the fatty acids in the form of soaps instead of producing biodiesel. Therefore, water is removed first regardless of the feedstock used.

(ii) Determination and treatment of free fatty acids

The concentration of the free fatty acids presence in the sample of vegetable oil is determine by do the titration of the cleaned feedstock oil with a standardized base solution. These acids are then either esterified into biodiesel, esterified into glycerides, or removed, typically through neutralization.

(iii) Reactions

Production of biodiesel with glycerol as the coproduct take place by base catalysed transesterification that reacts lipids, which are either fats or oils with short chain alcohol. Transesterification reaction by acid catalysed can also be used for production of biodiesel if the feedstock oil has high acid content.

The deprotonation of added alcohol with a base will take place in the transesterification reactions to make it a stronger nucleophile. In mild conditions, the reaction will either proceed very slowly or no reaction at all, therefore, the use of heat or catalysts are required to speed up the reactions. The acid or base added into the reactions are not consumed by the esterification reactions, thus it is named as the catalyst and not reactants. Sodium hydroxide, sodium methoxide and potassium hydroxide are the most common catalysts used in base catalysed transesterification reactions.

Base catalysed technique is used in almost all production of biodiesel from virgin vegetable oils. This is because, when treating virgin oils, only mild conditions (low temperature and pressure) is required with 98% conversion yield. Thus, it is the most economical process with low cost. However, acid catalysis are required in production of biodiesel from other sources and by other methods, which is much slower (Ataya et al. 2007). Only the base catalysed transesterification process will be described, since it is the predominant method for commercial scale production.

The transesterification reaction is base catalysed. Any strong base that have an ability to deprotonate the alcohol can act as the catalyst. However, sodium hydroxide (NaOH) and potassium hydroxide (KOH) is commonly used due to their low cost. The reaction must be kept dry, because any presence of water will result in undesirable base hydrolysis (Figure 6.2).

In the transesterification mechanism, the carbonyl carbon of the starting ester ($RCOOR_1$) undergoes nucleophilic attack by the incoming alkoxide (R_2O^-) to give a tetrahedral intermediate, which either reverts to the starting material, or proceeds to the transesterified product ($RCOOR_2$). The various species exist in equilibrium, and the product distribution depends on the relative energies of the reactant and product.

(iv) Product purification

The reaction will result in some by products such as glycerol, soap, excess alcohol and also water and not 100% biodiesel. However, to meet the standards, all the by-products generated must be removed, and the order of removal is process dependent. Difference in their properties is exploited in order to separate the bulk of the glycerol coproduct, where the density of biodiesel is smaller than that of the glycerol. Distillation process can recover the residue of methanol which will be reused in the other reactions. Residual water is removed from the fuel and conversion of soaps to acids also can removed the soaps. Homemade biodiesel is not recommended, even though the production process is relatively easy and simple. The homemade fuel might not meet the standards ASTM D6751 specifications that will result in engine damage, operational problems and also loss of warranty.

The main environmentally beneficial properties of biodiesel are that it was describes as 'carbon neutral' which means that no net output of carbon in the form of carbon dioxide (CO_2) will be produced. This is because, when the fuel is combusted, the amount of CO_2 released are the same as the amount of CO_2 absorbed when the oil crops grow. However, this theory is not completely accurate. It is because, the CO_2 that is released when the fertilizer is produced is exactly the same time the fertilizer is required to fertilize the fields at where the oil crops are grown. During production of biodiesel, there were many other sources that contribute to pollution other than the fertilizer production, such as the esterification process, drying, the solvent extraction of the oil and also transporting of the feedstocks and products.

All of these mentioned processes require an energy input weather in the form of fuel or electricity, in which both will results in the emission of greenhouse gases

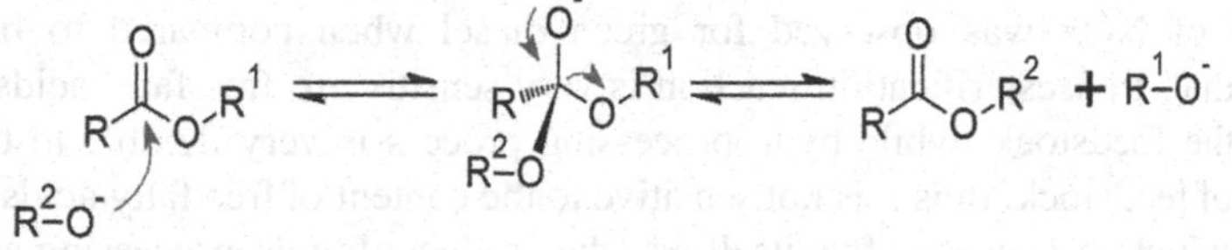

Fig. 6.2 General mechanism of transesterification reaction

that will contribute to pollution. Life cycle analysis technique are required in order to properly assess the impact of all these sources to environment. Fortunately, biodiesel is completely non-toxic and they are rapidly biodegradable. Thus, even if there are spillages of biodiesel, there are less of a risk than the spillage of fossil diesels. Besides that, the flash point of biodiesel is higher when compared to fossil diesel and so it is safer in the event of a crash.

6.2 Green Diesel Production

Green diesel is expected to be the next generation fuel for transportation which emerged because of the need for a renewable fuel replacement same as biodiesel. However, green diesel is better because it can be produced in large volumes at existing centralized petroleum refineries, unlike the biodiesel. In contrary, the biodiesel is more suitable with the smaller scale of production plants that usually situated in the rural areas where they are close to the source of oil used in the production process.

Hydroprocessing and decarboxylation/decarbonylation are the catalytic reactions that involves in the roduction of triglycerides from agricultural feedstocks to generate a mixture of diesel-lie hydrocarbons which are called as green diesel or renewable diesel (Hoekman 2009; Simakova et al. 2009). Reaction of triglycerides and free fatty acids (FFAs) to form water and n-paraffins by eliminates oxygen is carried out by hydrodeoxygenation reaction. While, the elimination of oxygen to form carbon dioxide or carbon monoxide and n-paraffins are carried out by decarboxylation or decarbonylation (Specht et al. 2005). Those reactions leads to a diesel product that is indistinguishable from petroleum diesel whereas biodiesel is chiefly composed of oxygenated species that can have vastly different properties than traditional petroleum diesel or fossil diesel (Smith et al. 2009). There is significant difference between biodiesel and green diesel even though they are both lipid-derived liquid transportation biofuels.

Their molecular structures are the first difference between the two fuels. The main constituents of green diesel are hydrocarbons, while, the biodiesel constituents are alkyl ester molecules. Therefore, no oxygen based molecules can be found in the green diesel, unlike biodiesel. Due to the absence of oxygenated molecules in the green diesel, they have higher heating value and higher energy density when compared to biodiesel (Lewis 2007). The second difference between these two fuels is that the biodiesel has cetane number on the order of 50, while an extremely higher cetane number (80–90) can be observed in green diesel. Thirdly, lower emission of NOx was observed for green diesel when compared to biodiesel. Besides that, transesterification reaction is very sensitive to free fatty acids (FFAs) level in the feedstock, while hydroprocessing process is very flexible to the composition of feedstock, thus it is not sensitive to the content of free fatty acids (FFAs). Propane which is a gaseous fuel itself was the product of hydroprocessing and it can be utilized in the system in regards to their side products (Hoekman 2009).

Furthermore, the hydrocarbons that was used as fuel have an outstanding energy density, and it makes them one of the option to be a powerful transportation fuel (Specht et al. 2005). Overall, when compared to biodiesel, green diesel seems to be more superior product based on the above arguments (Hoekman 2009). Therefore, the commercialization of green diesel has been started.

Currently, Neste Oil in Finland produced the green diesel industrially at two of their plants with a combined capacity of 170,000 ton/year (Simakova et al. 2009). In addition, they have started production of green diesel in Singapore and Rotterdam in 2010 and 2011 respectively with plants that have a capacity of 800,000 ton/year. UOP LLC and Eni cooperation also have led the commercialization of green diesel using vegetable oils. In 2009, the planned to use catalytic hydroprocessing technology in production of renewable diesel fuel have been started to get green diesel from vegetable oils by ecofining technology. A direct substitute for diesel fuel has been proposed by using products having a high cetane value (Kokossis and Yang 2010).

A hydrotreating process have been developed separately by Petrobas/H-BIO for the production of green diesel from vegetable oils and mineral diesel fractions, which then can be utilised as a diesel fuel cetane enhancer with additional benefit of reducing the density and also sulphur content (Kokossis and Yang 2010). Due to lack of process knowledge, the advancing in production of green diesel have been slow down. To further developed the current green diesel production process, a better understanding of the processing conditions is required. These processing conditions refer to catalyst composition, optimization of reaction conditions and also catalyst preconditioning/hydrotreating. In order for the production of green diesel is more competitive with production of petroleum diesel, the improvement in their process economics must be improved.

Production of green diesel via a catalytic reaction involves the process of either hydroprocessing or decarboxylation/decarbonylation of triglycerides from various agricultural feedstocks (Lewis 2007). Feedstocks that were derived from biomass usually contain oxygenated compounds that will results in lower chemical stability and energy content of the fuel. Therefore, to achieve a liquid fuel with higher thermal stability and combustion properties similar to petroleum fuels, the removal of oxygen from the feedstocks is a must. The process to remove oxygen from the feedstocks is called deoxygenation, and this deoxygenation process includes hydrodeoxygenation and decarboxylation/decarbonylation.

6.2.1   Hydrodeoxygenation (HDO) for Hydrogenation-Derived Renewable Diesel (HDRD)

Hydroprocessing is a catalytic reaction that was used to saturate the olefins and aromatics, and also to eliminate the heteroatoms such as nitrogen, sulphur, oxygen and metals by using hydrogen (Lewis 2007). Typical hydroprocessing reactions include hydrodenitrogenation (HDN), hydrodesulphurisation (HDS), hydrodeoxy-genation (HDO) and hydrodemetalisation (HDM). HDN is a reaction that was target to remove nitrogen as ammonia, HDS is to remove sulphur by breaking of C–S bond and followed by formation of hydrogen sulphide. HDO is carried out to remove oxygen as water, and lastly HDM is carried out to remove metal such as metal sulphide (Lewis 2007). The focus in production of hydrogenation-derived renewable diesel (HDRD) is to remove oxygen from the bio-oils or fats and it correspond to a HDO reaction. And the objective is to obtain hydrocarbon in diesel fuel range. Moderate temperatures (300–600 °C) and high hydrogen pressure are required to operate the HDO reactions of oil/fats in the presence of heterogeneous catalyst. However, the reactions conditions such as temperature and pressure should be adjusted depending on the feedstock (Lewis 2007). The same catalyst such as NiMo or CoMo are used in the HDO reactions, which are the same catalyst used in HDS and HDN. This is because, these hydrogenation processes are very similar in petroleum refineries (Fig. 6.3).

The production of green diesel as an alternative diesel fuel has been shown to be possible by the hydrogenation of triglycerides, which also named as HDRD. Due to its high cetane number, HDRD is preferred to be use as a diesel fuel additive providing a function to improve fuel ignition (Craig and Soveran 1991). In U.S. Pat. No. 4,992,605, Craig and Soveran had shown that the hydrocarbons in the diesel boiling range (C15–C18 paraffins) that can act as a fuel ignition improvers were produced in the hydroprocessing reactions of vegetable oil such as canola, soybean and sunflower. The process was carried out at a temperature in the range of 350–450 °C with a liquid hour space velocity (LHSV) of 0.5–5.0 per hour with the use of commercial catalyst such as NiMo and CoMo. The determination of optimum temperatures and pressures was carried out for selected vegetable oils as shown in Table 6.1.

In US Pat. No. 5,705,722, outline on another process for diesel fuel ignition improvers of green diesel production is discussed. According to Monnier et al., the production of HDRD with 80 wt% yield and cetane number of more than 90 is possible. A biomass feedstock was processed at temperature of 370 °C and

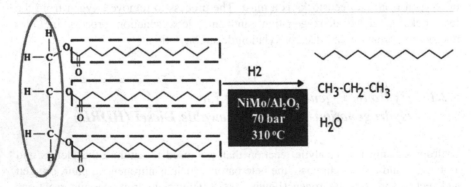

Fig. 6.3 The oxygen removal from the triglycerides (HDO reaction)

Table 6.1 Optimum hydroprocessing temperatures and pressures (Craig and Soveran 1991)

Feedstock	Canola oil	Sunflower oil	Soybean oil	Rapeseed oil	Tall oil fatty acid	Palm oil
Optimum temperature (°C)	370	360	360	390	390	370
Optimum pressure (MPa)	4.8	4.8	4.8	4.8	4.8	4.8

hydrogen pressure of 8.3 MPa for their work. A commercial catalyst NiMo/Al with silicon carbide (SiC) in a volume ratio of 2:1 was employed in the process. At the end of the processes, they conclude that better yield of cetane improvers can be obtained from hydrogenation of a mixture of tall oil with vegetable oil or animal fats when compared to a single feedstock. However, the use of high hydrogen pressures in the processes is not compatible with commercially feasible processes even though their yields and cetane numbers is very high.

6.2.2 Decarboxylation of Fatty Acids

A chemical reaction where a carboxyl group (COOH) is removed from a molecule as carbon dioxide is called as decarboxylation. While, decarbonylation is a chemical reaction where carbonyl group (C=O) is split off from a molecule. Using heat, decarboxylation of many carboxylic acids can be carried out by suspending the acid in an immiscible and high boiling-point liquid. Fatty acids are carboxylic acids; therefore, the same process can be applied to fatty acids to form straight hydrocarbon chains. When compared the decarboxylation reaction with hydrodeoxygenation reaction, there are several benefits of decarboxylation over hydrodeoxygenation even though they are both can successfully produce deoxygenated hydrocarbons (green diesel).

Hydrogen consumption in decarboxylation reaction is lower when compared to hydrodeoxygenation reaction because the hydrogen was requiring only to saturate the olefins. Meanwhile, for hydrodeoxygenation, the hydrogen is required not only to saturate the olefins, but also to remove oxygen as water. Due to less consumption of hydrogen, the reduction in size of the hydrogen purchases and hydrogen compressor leads to less capital and operational costs. In addition, because of the decarboxylation reaction is favoured at lower pressures when compared to hydrodeoxygenation reaction, the capital and operational costs are also lower (Lewis 2007). Another advantage of the decarboxylation process over hydrogenation is that, no formation of water will occur and it will result in better catalytic stability.

Even though additional production of carbon dioxide (CO_2) might occurs in both reactions, the CO_2 produced in decarboxylation reaction can be captured in a relatively pure state, thus it provides an additional advantage to the decarboxylation reaction over hydrodeoxygenation reaction. It has been shown that, when compounds of ethyl stearate or stearic acid were oxygenated using commercial catalyst (activated carbon supported palladium), the production of mainly n-heptadecane is possible (Kubičková et al. 2005). The intermediates in the decarboxylation reaction of stearic acid here is the heptadecenes and the reaction was carried out at the temperature of 300 °C and pressure of 17 bar. However, the best conversion efficiency (62%) in the decarboxylation reaction of stearic acid can be obtained when 5% of hydrogen and 95% of argon volumes were used as the reaction atmosphere when compared to the used of 100% helium (41%) or 100% hydrogen (49%). During deoxygenation, at first the conversion of ethyl stearate to stearic acid take place, and then will be further decarboxylated to n-heptadecane. At the same reaction conditions with decarboxylation of stearic acid, the best conversion yield of ethyl stearate decarboxylation was achieved. Aromatic groups are not desirable in diesel fuel, thus the selectivity to n-heptadecane is decreased because the aromatic groups start to form at 300–360 °C.

The reaction kinetics on decarboxylation of stearic acid and ethyl stearate for diesel fuel hydrocarbons production was studied in a semi-batch reactor over a catalyst (Pd/C) (Snåre et al. 2007). According to the study, by first order kinetics the conversion of ethyl stearate to stearic acid intermediates take place, which then by zero order reaction it will be converted to n-heptadecane at temperature of 300 °C. it was found that the deactivation of catalyst by decarboxylation reaction occur when the concentration of the intermediate product is high (Snåre et al. 2007). A study was carried out without the use of catalyst versus a set of different catalysts to get better understanding on the effects of heterogeneous catalyst on decarboxylation reaction. At the end of the study, conclusion was made; where only 5% of conversion of thermal decarboxylation without catalyst in a semi-batch reactor under helium atmosphere at temperature of 300 °C and pressure of 6 bar. Under the same reaction condition as without catalyst, a series of catalyst including carbon supported catalysts, metal oxides as well as Raney nickel catalyst were tried. From the results, it can be concluded that, most probably due to the metal support interaction, carbon supported catalysts lead to higher rates for decarboxylation reaction of stearic acid. When Ru/C and Rh/C is used as the catalyst, their deactivation might occur because their selectivity was higher towards the unsaturated side product (Kubičková et al. 2005). Another production of diesel fuel hydrocarbons by catalytic deoxygenation in a semi-batch reactor is studied using unsaturated fatty acids including monounsaturated fatty acids, oleic acid, linoleic acid, the di-unsaturated fatty acid, and the monounsaturated fatty acid ester, methyl oleate (Kubičková et al. 2005). Pd/C catalyst was employed and the reaction conditions used in this study is at temperature of 300–360 °C and pressure of 15–27 bar.

Later, another study for production of green diesel by decarboxylation of stearic acid is demonstrated using Pd catalyst supported on sibunit (a new class of mesoporous carbon-carbon composite materials combining advantages of chemical stability and electric conductivity of graphite and high specific surface area and adsorption capacity of active coals) (Lestari et al. 2008). This study was carried out in a semi-batch reactor at the temperature of 300 °C and at pressure of 17 bar with helium atmosphere and dodecane as a solvent. Under these conditions, the main products of n-pentadecane and n-heptadecane is formed via catalytic decarboxylation of stearic acid. Thus, it can be concluded that the type of support and the nature of the surface groups in carbon materials of catalyst influenced the distribution of products in decarboxylation of stearic acid (Lestari et al. 2008).

6.3 Fats and Oils from Renewable Sources

A natural resource which replenishes in order to overcome the depletion of resource caused by usage and consumption is called a renewable resource. The resource can be renewable either by naturally recurring processes in a finite amount of time in a human scale or through biological reproduction. Renewable resources is considered as the largest components of its ecosphere and the key indicator for a resources sustainability is the positive life cycle assessment (Stead 2015). Solar, wind power, biomass and geothermal are the common sources of renewable energy which are all categorised as renewable resources. Presently, the most important renewable feedstock of the chemical industry are fats and oils. In the production of oleochemical, tremendous geographical and feedstock shift from tallow to palm oil has taken place. The cultivation of more and new oil plants containing fatty acids which are desired for chemical utilization is very important while simultaneously the agricultural biodiversity will be increase. For industrial applications, which includes biodiesel, about 20% of the fats and oils global production were used which is about 160 million tonnes. The global production of biodiesel was developed widely and the problem of the industrial utilisation of food plant oils has become more urgent. Several reports have been discussed on the advances made using plant oils including organic synthesis, biotechnology and catalysis. For example, functionalization of fatty acids containing internal double bonds, application of the olefin metathesis reaction, and de novo synthesis of fatty acids from abundantly available renewable carbon sources. For the production of liquid biofuels for transportation, there is a large variety of biomass feedstock that are available (Pradhan et al. 2008) and it can be categorises into three basic categories as shown in Fig. 6.4 (Lewis 2007).

   A composite material of rigid cellulose fibers which are very large polymers that is composed of many glucose molecules is called lignocellulose. Lignin is a polymer constructed of alcohol units and non-carbohydrate; and hemicelluloses is a molecules consist of short, highly branched and sugar chains (Kauriinoja and Huuhtanen 2010). Diverse of lignocellulose materials can be obtained from different proportion of cellulose, hemicellulose and lignin in their composition. Typical biomass contains 10–25% lignin, 4–60% cellulose, and 20–40% hemicelluloses (Kauriinoja and Huuhtanen 2010). To be utilised in the production of nest generation biofuel, some of the suitable examples of the lignocellulosic feedstocks are listed in Fig. 6.4 (Kauriinoja and Huuhtanen 2010). Lignocellulosic

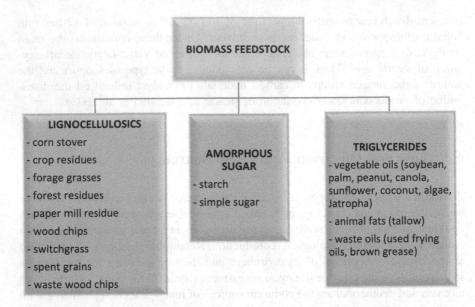

Fig. 6.4 Biomass feedstock classification for liquid transportation biofuel production

biomass is a feedstock with a low-energy-density, therefore, it is an expensive transportation fuel even though it is inexpensive and easy to find as a feedstock (Lewis 2007).

In order to produce liquid alternative biofuels, amorphous sugars such as simple sugars and starches also can be used as the biomass feedstocks. However, debate on fuel versus food is one of the main arguments against it (Malloy and Mayowski). Crop sources that can be used to feed people or livestock is made as the feedstock of biofuels generation is quite an issue. The idea on diverting crops away from food usage will create several adverse effects is supported by many people. Among these are the loss of farmable land to harvesting crops for starch and sugar based biofuels instead of food supplies for the population and livestock. In addition, the lack of crops availability for food consumption will results in price inflations of food items might occur if the crops is loss to production of fuel (Malloy and Mayowski; Muller et al. 2007).

The third group of biomasses that can be used as a feedstock for production of transportation fuels is triglycerides. Triglycerides are the main constituents that can be obtained from vegetable oils and animal fats. In the structure of triglycerides, glycerol can be found. Figure 6.5 shows the structure of triglyceride molecule and triacylglyceride of stearic acid consisting of three fatty acids and glycerol (Smith et al. 2009).

While, the structure of saturated and unsaturated fatty acid molecule was shown in Fig. 6.6. The structure of fatty acid molecule consist of the long straight aliphatic tails and carboxylic acids (Smith et al. 2009).

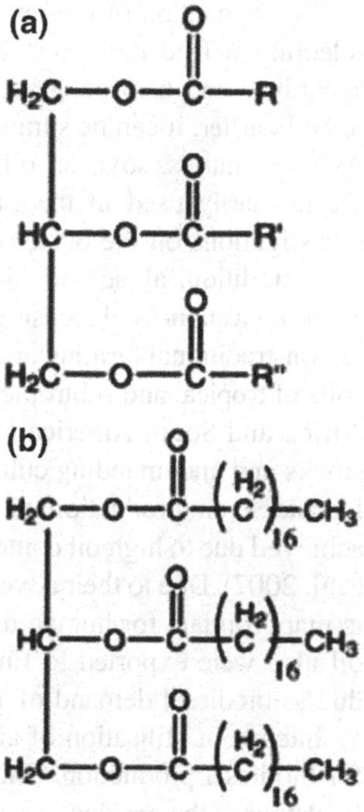

Fig. 6.5 a Structure of a
triglyceride molecule **b** the
triacylglyceride of stearic acid
(octadecanoic acid) (Smith
et al. 2009)

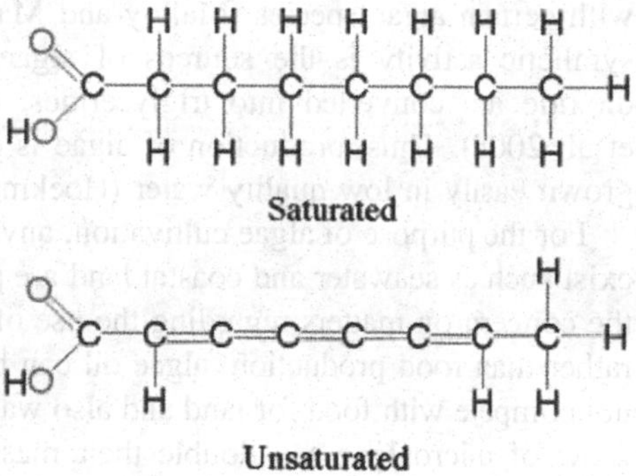

(a)

$$H_2C-O-\overset{\overset{\displaystyle O}{\|}}{C}-R$$

$$HC-O-\overset{\overset{\displaystyle O}{\|}}{C}-R'$$

$$H_2C-O-\overset{\overset{\displaystyle O}{\|}}{C}-R''$$

(b)

$$H_2C-O-\overset{\overset{\displaystyle O}{\|}}{C}-\left(\overset{H_2}{C}\right)_{16}-CH_3$$

$$HC-O-\overset{\overset{\displaystyle O}{\|}}{C}-\left(\overset{H_2}{C}\right)_{16}-CH_3$$

$$H_2C-O-\overset{\overset{\displaystyle O}{\|}}{C}-\left(\overset{H_2}{C}\right)_{16}-CH_3$$

Fig. 6.6 Saturated and
unsaturated fatty acids

Saturated

Unsaturated

All the three fatty acids in the structure of triglyceride can either be all the same
or different. Even numbered containing 16, 18 or 20 carbon atoms are the most
common length of the carbon chain (Smith et al. 2009). Typically, a mixture of fatty
acids can be found in triglycerides from vegetable oils (Lewis 2007). For example,
in soybean oil 7% of linolenic acid (with 3C=C), 51% of linoleic acid (with 2C=C),
10% of palmitic acid, 23% of oleic acid and 4% of stearic acid (Lewis 2007).
Soybean, palm, canola, peanut, coconut and sunflower are the most common
sources of vegetable oils.

For production of biofuel, some important considerations should be taken when selecting a feedstock such as local availability of the feedstock, financial manageability and also geography (Pradhan et al. 2008). Based on the criteria mentioned earlier, it can be summarized that; palm oil is the most common feedstock in Asian countries, soybean oil are preferred in the United States and sunflower oils are primarily used in the European Union (Pradhan et al. 2008). However, the investigations on the other seed oil also have been carried out (Lewis 2007).

In addition, algae and Jatropha are two of the new candidates that have been emerged as non-food feedstocks of biofuel because they can be harvested and grown in non-traditional farming areas (Nylund et al. 2008). Jatropha can grow well in poor soils of tropical and subtropical countries and they were grown mainly in Asia, India, Africa and South America (Nylund et al. 2008). High oil content, non-food feedstocks and undemanding cultivation are the advantages of Jatropha to be used as the feedstock on production of biofuel. Production of high-quality biofuels can be achieved due to high oil content of Jatropha seed which is about 27–40% oil (Achten et al. 2007). Due to their adverse effects associated with its consumption, Jatropha oil is inappropriate for human diets, therefor it is not a food based feedstock. Jatropha oil also were exported to Europe, where their domestic feedstocks are insufficient due to biodiesel demand of 10 Mt per year (Pickett et al. 2008).

Interest in utilisation of algal oils is growing nowadays to be use as a feedstock for biodiesel production. This is because, when compared to the other triglycerides feedstocks, the production of oil from algae is much higher (Schwietzke et al. 2008). Based on dry weight, the yield of oil from microalgae can be as high as 80% with certain algae species (Malloy and Mayowski). Varieties of microbes' photosynthetic activity is the sources of algae where sunlight, nitrogen and carbon dioxide are converted into triglycerides, carbohydrates, and lignin (Schwietzke et al. 2008). Thus, production of algae is easily manageable because they can be grown easily in low quality water (Hoekman 2009).

For the purpose of algae cultivation, anywhere conventional agriculture does not exist such as seawater and coastal land are preferred (Schwietzke et al. 2008). With the concern on matters regarding the use of agricultural land for energy generation rather than food production, algae oil can be the potential solution because it does not compete with food for land and also water resources. In addition, extremely fast grow of microalgae can double their mass in less than one day. While the first generation of biomass take up to 61% of farming land to supply the domestic fuel needs, using algae only 3% of the crop land are needed for the same purpose (Malloy and Mayowski). Recently, Shell and HR BioPetroleum have declared that they would use algae for the production of biodiesel.

According to HR BioPetroleum, when compared to the other terrestrial crops such as palm, rapeseed and soybean, the cultivation of algae yields nearly 15 times more oil per hectare (Schwietzke et al. 2008). In addition, news from a AlgaeLink and KLM regarding the cooperative effort on commercialization of algae was declared to use algae oil as the next generation alternative jet fuel for the operation of the Air France/KLM aircraft (Al Alwan 2014). AlgaeLink claimed that making the algae commercial is interesting for a large number of markets can be achieved, and they have been selling their systems since 2007 (Al Alwan 2014). Due to its low production cost, which is about 50 cents per gallon, algae oil have become more attractive to be use as an alternative fuel with alae as the feedstock (Nylund et al. 2008). However, algae oil has not been carried out extensively in a commercial scale even though they have numerous advantages when compared to the other vegetable oil resources (Hoekman 2009).

In addition, instead of vegetable oils, utilisation of animal fats as feedstocks also was used in biodiesel plant. In the production of biodiesel, the use of lard and fish oil as the fat resources have been demonstrated. The main feedstock for production of biodiesel in USA is soybean oil. However, since the price of the soybean oil have been increased, the biodiesel industry needs biomass resources that is much cheaper and it became a great driving force for them to use chicken fat as the feedstock for production of biodiesel. Tyson is one of the largest producer of leftover fat from chicken in US, and they announced that the production of animal fat was about 300 million gallons and it have a potential to be converted into fuel.

However, when the animal fats were used as the feedstocks, there will results in some technical drawbacks. This is because animal fats have a high cloud point which will limits their usage in the areas where the temperatures do not fall below 40 °F. Since the animal fats have high cloud point, they will cloud up more at high temperatures than soy based biodiesel and also thickens when used in colder climates.

Vegetable oil and animal fats are not the only constituents of triglycerides feedstock; there is also waste oils such as brown grease and used frying oils. Currently, waste cooking oils is a very important source for economical production-oriented approaches since they are the lipid feedstocks with lower cost. However, difficulty in processing of the feedstocks is the main issues due to their inconsistencies in the composition of the oil (Pradhan et al. 2008). To obtain biodiesel with high grade, the alteration to the conversion method have to be done depends on the variations in water content, triglycerides, impurities and also composition of free fatty acids (FFAs) (Pradhan et al. 2008). In fact, variables in quality of the waste oil as feedstock is more problematic when compared to the vegetable oils (Nylund et al. 2008).

Important consideration should be taken to choose the biomass that need to be used as the source of biofuels production (Smith et al. 2009). Its chemical and physical characteristics of the feedstock, as well as supply, storage properties, cost and also engine performance also need to be considered in order to determine if it is suitable for commercial production of biofuel or not (Pradhan et al. 2008). Relatively high cost of the triglyceride feedstock is the main reason of the economic defeats of biofuels against petroleum based fuels. About 70–85% of the total production cost is related to the cost of feedstocks (Pradhan et al. 2008). Currently, the lowest cost on production of biodiesel is by using waste oil and animal grease as the feedstock together with available technologies (Pickett et al. 2008). With respect to other available biomass feedstocks, generally, the cheapest feedstocks are lignocellulosic, followed by amorphous sugars, and the most expensive feedstocks are triglyceride (Lewis 2007).

6.4   Catalytic Upgrading of Fats and Oils from Renewable Sources

The design of catalyst systems to be used in the processing of renewable materials needs to be rational and several factors should be taken into consideration. The most important thing is that under rough conditions inside the reactor, the catalyst can still work well. The rough conditions inside the reactors occurs as the results of formation of carbon monoxide (CO) that will inhibit the desulphurisation. The catalyst also must have an ability to handle the increase in the consumption of hydrogen and also a large increase in temperature due to the fast reactions in the reactor. Furthermore, the issues with the poor cold flow properties of the products due to high content of n-paraffins also need to be solved properly. Many issues will be arising in the processing reactor if the catalyst used is not designed or tailor-made to handle the specific function, and it depends on the amount and quality of the materials that are blended into the pool of diesel feed. Some examples of the arising issues with the used of not designed catalyst are hydrogen starvation, build-up of pressure drop, poor desulphurisation and lastly the cold flow properties of the products might not meet the requirement. Therefore, evaluation need to be done carefully on the catalyst design used for the hydrotreater treating biofuel.

Because of some reasons related to environment, the biomass has been the focused as a renewable source of chemicals and also fuels for the past decades. Biodiesel is a fuel that is derived from renewable sources such as vegetable oils and also animal fats (Guo et al. 2006). Attentions have been given to the biodiesel to be used as a direct replacement for the petroleum based diesel (Lestari et al. 2009). Transesterification reaction is one of the methods that is commonly use in the production of biodiesel. Even though this transesterification reaction might be driven by supercritical conditions, commonly it was carried out in the presence of catalyst. The products of transesterification reactions such as fatty acid methyl ester (FAME) and fatty acid ethyl ester (FAEE) can be used directly as the biodiesel or be blended with petroleum-based diesel.

6.4.1   Homogeneous Trans-esterification Catalysts

One of the advantage for homogeneous catalyst or base catalysis is that it can increase the reaction rates because of their constant contact with the reaction mixture. However, the loss of raw material occurred in neutralisation step which is required in order to remove the catalyst. Miao and Wu reported on the production of biodiesel through acid catalysed transesterification using microalgal oil obtained from C. protothecoides. The preliminary test using methanol and sulphuric acid suggested that the use of alkali catalyst in transesterification reaction is not suitable. It can be due to the high acid value of the microalgal oil (Miao and Wu 2006). It was found out that, the best combination of reaction conditions to be used in transesterification reaction was with 100% catalyst, molar ratio of 56:1 for methanol to oil, temperature at 30 °C and the reaction time of 4 h.

Sodium hydroxide, sodium methoxide, potassium hydroxide and potassium methoxide are the most common homogeneous catalysts that have been used for transesterification reaction, and their comparison have been done by Vicente el. al. with the reaction condition of temperature at 65 °C and molar ratio of methanol to sunflower oil is 6:1; the highest yield which is about 99.3% of the biodiesel can be achieved with the use of sodium methoxide with 99.7% purity, followed by potassium methoxide, potassium hydroxide and sodium hydroxide, with yields of 98.5, 91.7 and 86.7% respectively. The order reported was in a good agreement with the results of transesterification reaction of used oil by using sodium methoxide as the catalyst with the yield of 89%, which is higher when compared to yields obtained when sodium hydroxide and potassium hydroxide were used (Leung and Guo 2006).

The amounts of free fatty acids (FFAs) in plant and algal oil is high. Separation problems in the production downstream can occur if these are nor pre-treated because during transesterification they can react with the homogeneous base catalyst and form the corresponding soaps (Huber et al. 2006). The cost on production of biodiesel can be reduce by the use of lower grade feedstocks, however, the compositions of that feedstocks have higher numbers of free fatty acids (FFAs) (Canakci and Sanli 2008). When reaction between water and fatty acid methyl ester (FAME) occurred, the FFAs can also be formed. Another feedstock that can be used in production of biodiesel which is relatively cheap is greases, however they also contain high number of FFAs and without pre-treatment it will lead to some problems in the base-catalysed transesterification process. To carry out the stage of pre-treatment, the cost of production will be increase because they require homogeneous acid-catalyst which needs to be neutralised before the esterification reaction. The usage of immobilised diphenylammonium heterogeneous catalyst have been investigated by Zafiropolous et al. and they are successfully reduce the amount of FFAs from 11 to 1% in a pre-treatment process of greases (Chakraborty et al. 2010). After the process of the pre-treatment, base-catalysed transesterification reaction is carried out with the use of sodium methoxide as the catalyst at temperature of 50 °C for 2 h and gave high yield of FAME which is 99% (Haas et al. 2002).

6.4.2   Heterogeneous Trans-esterification Catalysts

Unlike the homogeneous catalysts, heterogeneous catalyst is more efficient and they provide benefit in the aspect of economy in production of biofuels. Heterogeneous catalysts were readily recycled after the transesterification reactions because they can be separated easily, thus the production costs can be lowered. Layered double hydroxides (LDHs) is a compound which consist of positively charged layered materials that have charge balancing anions in their interlayer region and this compound is considered as promising compounds to be used in the transesterification reactions. Choudary et al. have reported that, a wide range of anions which can catalyse a wide range of transesterification reactions can be intercalated, such as tert-butoxide, (Choudary et al. 2000) which includes the production of emulsifiers for food products (Corma et al. 1998). The formation of LDH materials consisting of hydrotalcite can be done by co-precipitation of soluble metal salts (Cantrell et al. 2005).

Depends on the anion intercalated into the layers and the ratio of Al:Mg, the properties of the solid base materials can be tuned. To avoid the contamination of alkaline in the catalyst, the alkali free method of co-precipitation can be used. The transesterification reaction of glyceryl tributyrate, hexyl ether and methanol was carried out in a stirred batch reactor using LDHs, with the reaction conditions at temperature of 60 °C for 3 h with 0.05 g of calcinated catalyst. And then, using gas chromatography (GC) the reaction was periodically sampled. The conversion of glyceryl tributyrate into methyl butanoate was carried out by di-glycerides and mono-glycerides. Due to the increase in Mg content and intralayer electron density, highest conversion of hydrotalcite can be led using ratio of Mg to Al of 2.93:1. Other bivalent and trivalent metal ions can be substituted partially or completely in the layered structure using both Mg^{2+} and Al^{3+} respectively. The formation of mixed metal oxides (MMOs) which is more basic than the corresponding parent layered samples can be obtained by the calcination process of LDH materials (Cavani et al. 1991).

In a recent study, the production of biodiesel were tested using LDH samples which have been calcined and were doped using various metal ions in order to replace Al^{3+} (Macala et al. 2008). The conversion of triacetin to the corresponding methyl esters was increased to around 80% with the used of 10% Gallium dopants at 60 °C. even greater yield which is more than 95% can be achieved with the use of 5–20% of Fe based dopant when compared to 1.0% weight of dopant at the temperature of 60 °C for 40 min. This catalyst has a surface area which is greater by 50% than the uncalcined MgAl hydrotalcite. Significantly higher surface area of the mixed oxides derived from LDHs that was calcined at temperatures of 400–550 °C when compared to the samples of parent LDH. Only 50% of the initial activity of original catalyst can be obtained when catalyst was re-calcined and extracted during regeneration via rehydration. The next regeneration step also will yield the same results.

Bo et al. has reported a process on conversion of palm oil to alkyl esters with the utilisation of KF/Al_2O_3 as the solid-base catalyst (Bo et al. 2007). The preparation of catalyst involved the impregnation of the KF in order to give the Al_2O_3 a supported catalyst. The catalyst was then dried and followed by calcination at temperature of 600 °C. At an optimum temperature of 65 °C and atmospheric pressure condition, the transesterification reaction was carried out. At the temperature which is higher than that optimum temperature, the decrease in desired ratio of methanol to oil (12:1) will occurs as the results of volatility of methanol. The production of triglyceride with the yields of over 90% can be achieved when 4% of catalyst with ratio of $KF:Al_2O_3$ is 0.331 carried out for over 4 h. Interestingly, new phase of K_3AlF_6 was produced by do the calcination of the catalyst at 600 °C and they were characterised by using the Thermogravimetric Analysis (TGA) and X-ray diffraction (XRD). Calcination of $Eu(NO_3)_3/Al_2O_3$ in different conditions was carried out; 2 h at temperature of 550 °C, 2 h at temperature of 300 °C and 8 h at temperature of 900 °C in order to prepare a superbase by forming Eu_2O_3/Al_2O_3 with an optimal Eu content of >6.75% (Li et al. 2007). The superbase will then be used in a fixed bed reactor at atmospheric pressure to do the transesterification of soybean oil. And again, due to the volatility of methanol, the optimum temperature for the reaction was at around 70 °C.

In order to prevent the reaction of water with the catalyst, removal of water from the oil and methanol is carried out. When observed by gas chromatography (GC), for the first 30 min no reaction was observed, started from 2 h reaction time steady increase in the rate of reaction take place, and finally at 8 h reaction time the final conversion is up to 63%. The reduction on catalyst activity will occur after it have been used for 40 h and it will result in conversion for only 35%, might be due to the presence of water and the free fatty acids (FFAs). The decreased in its activity and the loss of surface area will be occurred after each subsequent regeneration of catalyst. For the purpose of biodiesel production, some potential oils such as deep frying oil can be used. However, high content of FFAs in these oil makes them not suitable for base catalysed transesterification reaction. In that case, the most preferred catalyst is a heterogeneous acid catalyst. For example, the production of biodiesel from soybean oil have successfully been carried out with 98.6% FAME yield when they was catalysed by sulphated zirconia ($SZrO_2$) catalyst (Garcia et al. 2008).

However, the deactivation of the catalyst occurs rapidly. Conversion of waste cooking oil that contains 15% of FFAs to 98% yield of FAME was carried out at relatively high temperature (200 °C) using zinc stearate that was immobilised on silica gel results in no loss of activity even after four catalytic cycles (Smith et al. 2009). Transesterification of oils with 27.8% of FFAs with the presence of carbohydrate-derived heterogeneous acid catalysts have been successfully converts the oils into FAME after 8 h with 92% yield (Lou et al. 2008). Even after 50 successive uses, there are still about 93% of these catalysts were still active and stable. Fuels with different properties can be formed by vary the alcohol used in the transesterification reaction. Methanol is usually chosen to be used in the transesterification reaction. Bokade et al. has reported percentage conversion using different alcohols varied from methanol to n-octanol with the use of TPA/K-10 as the catalyst; methanol (84%), ethanol (80%), n-propanol (76%) and n-octanol (72%). From the results reported, it shows that the percentage of conversion were decrease when the number of carbon atoms were increase. The increase in number of carbon atom might results in slow down of the rate of reaction. In order to prevent the waste of energy in the transesterification reaction, it is very important to know when the process has reached completion and how far the reaction has progressed along its reaction profile. Measurements of in situ viscosity using an acoustic wave viscometer have been tested in order to monitor the progress of the transesterification process. In the viscosity measurement, a characteristic plateau will be observed

Table 6.2 Selected reactions for trans-esterification of biomass lipid to biofuel conversion

Author	Year	Catalyst	Reaction conditions	Product	Reported yields (%)
Vicente et al. (2004)	2004	Sodium hydroxide	65 °C, 6:1 methanol:oil 1% catalyst (wt%) t ¼ 4 h	FAME	86.7
Vicente et al. (2004)	2004	Potassium hydroxide	65 °C, 6:1 methanol:oil 1% catalyst (wt%) t ¼ 4 h	FAME	91.7
Vicente et al. (2004)	2004	Sodium methoxide	65 °C, 6: 1 methanol:oil 1% catalyst (wt%) t ¼ 4 h	FAME	99.3
Vicente et al. (2004)	2004	Potassium methoxide	65 °C, 6:1 methanol:oil 1% catalyst (wt%) t ¼ 4 h	FAME	98.5
Bo et al. (2007)	2007	KF/Al$_2$O$_3$ 0.331 (wt/wt)	65 °C, 12:1 methanol:oil 4% calcined catalyst (wt%) t ¼ 3 h.	FAME	>90
Li et al. (2007)	2007	Eu$_2$O$_3$/Al$_2$O$_3$ with Eu 0.45–9.00 wt%	70 °C, 6:1 methanol oil. 4% calcined catalyst (wt%) t ¼ 8 h	FAME	63
Bokade and Yadav (2007)	2007	TPA/K-10 (dodecatungstophosphoric acid)	170 °C, 5:1 methanol:oil. 10% TPA/K-10 catalyst (wt%) t ¼ 8 h	FAME	84
Azcan and Danisman (2008)	2008	Potassium hydroxide	50 °C, 6:1 methanol:oil, 1% catalyst (wt%), microwave heating, t ¼ 5 min	FAME	93.7
Azcan and Danisman (2008)	2008	Sodium hydroxide	40 °C, 6:1 methanol:oil. 1% catalyst (wt%), microwave heating, t ¼ 3 min	FAME	92.7
Barakos et al. (2008)	2008	MgA–CO$_3$ hydrotalcite layered double hydroxide	200 °C, 6:1 methanol:oil. 1% catalyst (wt%), high initial FFA content, t ¼ 3 h	FAME	99 (final FFA < 1%)
Kouzu et al. (2008)	2008	CaO	Reflux, 12:1 methanol:oil. 14 mmol catalyst, t ¼ 1 h	FAME	93

(continued)

Table 6.2 (continued)

Author	Year	Catalyst	Reaction conditions	Product	Reported yields (%)
Lou et al. (2008)	2008	Starch-derived catalyst	200 °C, 30:1 methanol:oil. 10% catalyst (wt%) t ¼ 8 h	FAME	92
Garcia et al. (2008)	2008	S–ZrO₂	120 °C, 20:1 methanol:oil. 5% catalyst (wt%) t ¼ 1 h	FAME	98.6
Saydut et al. (2008)	2008	Sodium hydroxide	60 °C, 6:1 methanol:oil. 0.5% catalyst (wt%) t ¼ 2 h	FAME	74
da Silva et al. (2008)	2008	Co(II) ions adsorbed in chitosan	70 °C, 6: 1 methanol: oil. 2% catalyst (wt%) t ¼ 3 h, pH 8.5	FAME	94

once the reaction has been completed. In both bench-top batch scale and also plant scale, the measurements were successfully achieved. By using these progression measurements, the efficiency can be increase and the productivity can be maximised, at the same time results in biodiesel ever more competitive as a source of fuel.

6.4.3   Novel Energy Sources (Microwaves) for Trans-esterification

The microwave heating method also can be used to increase the rate of the transesterification reactions. The process of methanol transesterification of rapeseed oil have been carried out using optimised temperature of 60 °C and reaction time of 5 min by using the potassium hydroxide as the catalyst (Azcan and Danisman 2008). When compared to the other reactions without microwaves, this reaction is significantly quicker with the yields of 93.7% biodiesel and 97.8% of purity (Makareviciene and Janulis 2003). By using the same microwave methods and sodium hydroxide as the catalyst, production of biodiesel was carried out at the temperature of 40 °C for 3 min and results in 90.9% yield of biodiesel with 93.7% of purity. An overview of some transesterification reactions for biofuel production that have been carried out are shown in the Table 6.2.

Currently, many organisations around the world doing research in production of high grade renewable fuels of sustainable origin in both small and large scales. Breakthroughs in heterogeneous catalysis for decarboxylation reactions that allow the utilisation of existing petrochemical processing and refining infrastructure, and generate a product that is compatible with existing transport are likely to make biofuels a viable alternative energy source.

References

Achten WM, Mathijs E, Verchot L, Singh VP, Aerts R, Muys B (2007) Jatropha biodieheterogeneous catalyst for transesterificationsel fueling sustainability? Biofuels Bioprod Biorefin 1 (4):283–291

Al Alwan BA (2014) Biofuels production via catalytic hydrocracking of Ddgs corn oil and hydrothermal decarboxylation of oleic acid over transition metal carbides supported on Al-Sba-15

Anastopoulos G, Zannikou Y, Stournas S, Kalligeros S (2009) Transesterification of vegetable oils with ethanol and characterization of the key fuel properties of ethyl esters. Energies 2(2): 362–376

Ataya F, Dubé MA, Ternan M (2007) Acid-catalyzed transesterification of canola oil to biodiesel under single-and two-phase reaction conditions. Energy Fuels 21(4):2450–2459

Azcan N, Danisman A (2008) Microwave assisted transesterification of rapeseed oil. Fuel 87(10):1781–1788

Barakos N, Pasias S, Papayannakos N (2008) Transesterification of triglycerides in high and low quality oil feeds over an HT2 hydrotalcite catalyst. Biores Technol 99(11):5037–5042

Bo X, Guomin X, Lingfeng C, Ruiping W, Lijing G (2007) Transesterification of palm oil with methanol to biodiesel over a KF/Al₂O₃ heterogeneous base catalyst. Energy Fuels 21(6): 3109–3112

Bokade V, Yadav G (2007) Synthesis of bio-diesel and bio-lubricant by transesterification of vegetable oil with lower and higher alcohols over heteropolyacids supported by clay (K-10). Process Saf Environ Prot 85(5):372–377

Canakci M, Sanli H (2008) Biodiesel production from various feedstocks and their effects on the fuel properties. J Ind Microbiol Biotechnol 35(5):431–441

Cantrell DG, Gillie LJ, Lee AF, Wilson K (2005) Structure-reactivity correlations in MgAl hydrotalcite catalysts for biodiesel synthesis. Appl Catal A 287(2):183–190

Cavani F, Trifirò F, Vaccari A (1991) Hydrotalcite-type anionic clays: preparation, properties and applications. Catal Today 11(2):173–301

Chakraborty R, Bepari S, Banerjee A (2010) Transesterification of soybean oil catalyzed by fly ash and egg shell derived solid catalysts. Chem Eng J 165(3):798–805

Choudary B, Kantam ML, Reddy CV, Aranganathan S, Santhi PL, Figueras F (2000) Mg–Al–O–t-Bu hydrotalcite: a new and efficient heterogeneous catalyst for transesterification. J Mol Catal A Chem 159(2):411–416

Corma A, Iborra S, Miquel S, Primo J (1998) Catalysts for the production of fine chemicals: production of food emulsifiers, monoglycerides, by glycerolysis of fats with solid base catalysts. J Catal 173(2):315–321

Craig WK, Soveran DW (1991) Production of hydrocarbons with a relatively high cetane rating. Google Patents

da Silva RB, Neto AFL, dos Santos LSS, de Oliveira Lima JR, Chaves MH, dos Santos JR, de Lima GM, de Moura EM, de Moura CVR (2008) Catalysts of Cu (II) and Co (II) ions adsorbed in chitosan used in transesterification of soy bean and babassu oils–a new route for biodiesel syntheses. Biores Technol 99(15):6793–6798

Garcia CM, Teixeira S, Marciniuk LL, Schuchardt U (2008) Transesterification of soybean oil catalyzed by sulfated zirconia. Biores Technol 99(14):6608–6613

Guo H, Zhu G, Li H, Zou X, Yin X, Yang W, Qiu S, Xu R (2006) Hierarchical growth of large-scale ordered zeolite silicalite-1 membranes with high permeability and selectivity for recycling CO₂. Angew Chem 118(42):7211–7214

Haas MJ, Bloomer S, Scott K (2002) Process for the production of fatty acid alkyl esters. Google Patents

Hoekman SK (2009) Biofuels in the US–challenges and opportunities. Renewable Energy 34 (1):14–22

Huber GW, Iborra S, Corma A (2006) Synthesis of transportation fuels from biomass: chemistry, catalysts, and engineering. Chem Rev 106(9):4044–4098

Kauriinoja A, Huuhtanen M (2010) Small-scale biomass-to-energy solutions for northern periphery areas. MSc Thesis, Oulu

Kokossis AC, Yang A (2010) On the use of systems technologies and a systematic approach for the synthesis and the design of future biorefineries. Comput Chem Eng 34(9):1397–1405

Kouzu M, Kasuno T, Tajika M, Yamanaka S, Hidaka J (2008) Active phase of calcium oxide used as solid base catalyst for transesterification of soybean oil with refluxing methanol. Appl Catal A 334(1):357–365

Kubičková I, Snåre M, Eränen K, Mäki-Arvela P, Murzin DY (2005) Hydrocarbons for diesel fuel via decarboxylation of vegetable oils. Catal Today 106(1):197–200

Lestari S, Mäki-Arvela P, Beltramini J, Lu G, Murzin DY (2009) Transforming triglycerides and fatty acids into biofuels. ChemSusChem 2(12):1109–1119

Lestari S, Simakova I, Tokarev A, Mäki-Arvela P, Eränen K, Murzin DY (2008) Synthesis of biodiesel via deoxygenation of stearic acid over supported Pd/C catalyst. Catal Lett 122(3–4): 247–251

Leung D, Guo Y (2006) Transesterification of neat and used frying oil: optimization for biodiesel production. Fuel Process Technol 87(10):883–890

Lewis NS (2007) Toward cost-effective solar energy use. Science 315(5813):798–801

Li X, Lu G, Guo Y, Guo Y, Wang Y, Zhang Z, Liu X, Wang Y (2007) A novel solid superbase of Eu_2O_3/Al_2O_3 and its catalytic performance for the transesterification of soybean oil to biodiesel. Catal Commun 8(12):1969–1972

Lou W-Y, Zong M-H, Duan Z-Q (2008) Efficient production of biodiesel from high free fatty acid-containing waste oils using various carbohydrate-derived solid acid catalysts. Biores Technol 99(18):8752–8758

Macala GS, Robertson AW, Johnson CL, Day ZB, Lewis RS, White MG, Iretskii AV, Ford PC (2008) Transesterification catalysts from iron doped hydrotalcite-like precursors: solid bases for biodiesel production. Catal Lett 122(3–4):205–209

Makareviciene V, Janulis P (2003) Environmental effect of rapeseed oil ethyl ester. Renewable Energy 28(15):2395–2403

Malloy D, Mayowski K Biofuels for a sustainable world

Miao X, Wu Q (2006) Biodiesel production from heterotrophic microalgal oil. Biores Technol 97 (6):841–846

Muller M, Yelden T, Schoonover H (2007) Food versus fuel in the United States: can both win in the era of ethanol? Institute for Agriculture and Trade Policy (IATP) Environment and Agriculture Program Minneapolis, MN

Nylund N-O, Aakko-Saksa P, Sipilä K (2008) Status and outlook for biofuels, other alternative fuels and new vehicles, vol VTT-TIED-2426, VTT

Olivares RDC, Rivera SS, Mc Leod JEN (2014) Database for accidents and incidents in the biodiesel industry. J Loss Prev Process Ind 29:245–261

Pickett J, Anderson D, Bowles D, Bridgwater T, Jarvis P, Mortimer N, Poliakoff M, Woods J (2008) Sustainable biofuels: prospects and challenges. The Royal Society, London

Pradhan A, Shrestha DS, Gerpen JV, Duffield J (2008) The energy balance of soybean oil biodiesel production: a review of past studies. Trans ASAE (Am Soc Agric Eng) 51(1):185

Saydut A, Duz MZ, Kaya C, Kafadar AB, Hamamci C (2008) Transesterified sesame (*Sesamum indicum* L.) seed oil as a biodiesel fuel. Biores Technol 99(14):6656–6660

Schwietzke S, Ladisch M, Russo L, Kwant K, Mäkinen T, Kavalov B, Maniatis K, Zwart R, Shahanan G, Sipila K (2008) Analysis and identification of gaps in the research for the production of second-generation liquid transportation biofuels. Energy Research Centre of the Netherlands ECN

Simakova I, Simakova O, Mäki-Arvela P, Simakov A, Estrada M, Murzin DY (2009) Deoxygenation of palmitic and stearic acid over supported Pd catalysts: effect of metal dispersion. Appl Catal A 355(1):100–108

Smith B, Greenwell HC, Whiting A (2009) Catalytic upgrading of tri-glycerides and fatty acids to transport biofuels. Energy Environ Sci 2(3):262–271

Snåre M, Kubičková I, Mäki-Arvela P, Eränen K, Wärnå J, Murzin DY (2007) Production of diesel fuel from renewable feeds: kinetics of ethyl stearate decarboxylation. Chem Eng J 134(1):29–34

Specht M, Zuberbuhler U, Bandi A (2005) Why biofuels?–an introduction into the topic. RENEW–1st European summer school on renewable motor fuels, Uwelt-Campus-Birkenfeld, 29th–31st August

Stead JG (2015) Management for a small planet. ME Sharpe

Vicente G, Martınez M, Aracil J (2004) Integrated biodiesel production: a comparison of different homogeneous catalysts systems. Biores Technol 92(3):297–305

Chapter 7
Catalytic Upgrading of Glycerol, Conversion of Biomass Derived Carbohydrates to Fuels and Catalysis in Depolymerization of Lignin

7.1 Catalytic Upgrading of Glycerol

Over the past decades, petroleum resources have been used dominantly in production of petrochemicals and also fuels for transportation results in the depletion of petroleum resources. The consumption of these resources also has been considered as a main contribution for adverse climate change and also global environmental issues. Due to these issues, development and investigation on alternatives method for production of chemicals and fuels have been carried out by using biomass or biomass-derived compounds as the feedstocks. Biomass and biomass-derived compounds such as sorbitol, lactic acid, glucose and also glycerol (1,2,3-propanetriol) are renewable resources that can be used in production processes of chemicals and fuels (Zhou et al. 2011; Carlson et al. 2009; Bozell and Petersen 2010; Fan et al. 2009). Due to their reactivity, abundancy and also applicability, glycerol has been the most common feedstocks used as alternate routes in the past decades (Bozell and Petersen 2010; Towey et al. 2011; Brandner et al. 2009; Sharma et al. 2008; Gallezot 2012).

Glycerol is the building blocks of triglycerides which commonly exists in both animal fats and vegetable oils (Ilyushin et al. 2008). Additionally, glycerol molecules also exists intracellularly in algae with considerable quantity (Kaçka and Dönmez 2008; Hadi et al. 2008). The production of glycerol can also be done through catalytic composition and also catalytic hydrogenolysis (Arroyo-López et al. 2010; Modig et al. 2007; Seo et al. 2009; Wang et al. 2001). In essence, the origins of all these bio-resources is carbon dioxide and also water that produced by photosynthesis as shown in Fig. 7.1.

In industry, for soap production, as shown in Fig. 7.2a triglyceride is used in saponification reaction (Israel et al. 2008) and for fatty acid methyl esters (FAME) which also called as biodiesel production, triglyceride is used in transesterification reaction as shown in Fig. 7.2b (Saleh et al. 2010). In these reactions, the production of glycerol as by-product is unavoidable.

© Springer International Publishing AG 2017
S. Bagheri, *Catalysis for Green Energy and Technology*,
Green Energy and Technology, DOI 10.1007/978-3-319-43104-8_7

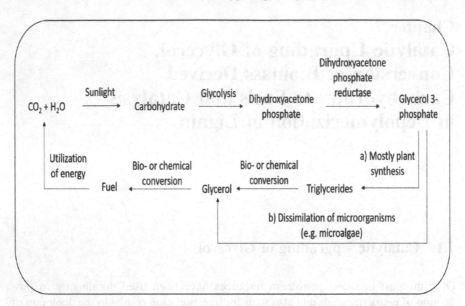

Fig. 7.1 Bioavailability of glycerol. **a** The metabolism of the plant yielding triglycerides which can be readily converted to glycerol. **b** The formation of glycerol by dissimilation of microorganisms

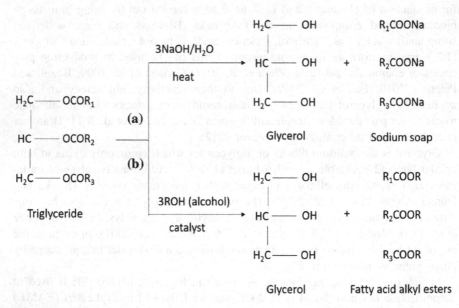

Fig. 7.2 Glycerol as a by-product in the reactions of **a** saponification of triglycerides and **b** transesterification of triglycerides with alcohol

As a result, the production of glycerol by traditional petrochemical methods have been uneconomic (Wang et al. 2001). Particularly, in some countries, the commercialization of biodiesel production has been witnessed in the past two decades. Practically, fatty acid methyl ester (FAME) can be used in the existing diesel engines after the production of FAME is done by transesterification of triglycerides with alcohol (glycerol produced as by-product) in the presence of alkaline catalyst and then mixed with petroleum-based diesel (Johnson and Taconi 2007; Tan et al. 2010). Approximately, for every tonne of FAME or biodiesel products around 0.1 tonne of crude glycerol is produced simultaneously as by-product in the transesterification reaction between methanol and triglycerides (Yaakob et al. 2013; Jerzykiewicz et al. 2009). Therefore, the use of such crude glycerol economically is required for every process of biodiesel production (Manosak et al. 2011; López et al. 2009; Devi et al. 2009; Ayoub and Abdullah 2012). Otherwise, the economics of biodiesel industry will be affected negatively and severe disposal problem might occur.

The products of transesterification between methanol and triglycerides in a homogeneously catalytic process usually will present themselves in two layers which are crude glycerol layer and also FAME layer (Cydzik-Kwiatkowska et al. 2010; Leung et al. 2010). Typically, at this stage, the purity of crude oil separated was about 80%, with another 20% of contaminants such as traces of glycerides and alcohol, water, ester and also organic/inorganic salts (Hájek and Skopal 2010). However, to be used as a reactant, pure glycerol is needed in most of the catalytic reactions that will be carried out for conversion of glycerol to any value-added products. Therefore, further refining and upgrading processes are required for the crude glycerol fraction in the process of biodiesel. These upgrading and refining processes are important in order to remove any residual of water, methanol, organic matter and also salts. Usually, to yield glycerol of industrial-grade with 90% of purity (Lakshmi Ch et al. 2009), processes like extraction, filtration (Manosak et al. 2011; Tizvar et al. 2009; Lakshmi Ch et al. 2009; Ye et al. 2011) and distillation (Kongjao et al. 2010) were used. The trace impurities were then further removed by processes such as deodorizing, bleaching and also ion exchange treatments in order to yield high purity glycerol up to 99.9% (Lakshmi Ch et al. 2009; Carmona et al. 2009). The refinery process of the crude glycerol can be simplified by the use of a heterogeneous catalyst in the production of biodiesel. For instance, a continuous production of biodiesel via transesterification reaction catalysed by heterogeneous catalyst was reported by Institut Français du Pétrole (IFP) in France. Remarkably, there is no aqueous treatment or catalyst recovery steps are required in that heterogeneous process. In addition to a high yield of biodiesel in each process, crude glycerol with no contaminant of salts and purity which is greater than 98% can be achieved directly (Bournay et al. 2005). The finding shows that the cost-intensive and tedious purification processes of crude glycerol can be avoided once a heterogeneous process was properly developed (Santacesaria et al. 2012; Atadashi et al. 2012).

The production of synthesis gas (syngas) or hydrogen can be carried out by using glycerol through catalytic partial oxidation with O_2 (Wang 2010), reforming with CO_2 (Wang et al. 2009; Fernández et al. 2010) or H_2O (Adhikari et al. 2008; Pompeo et al. 2010) pyrolysis (Kale and Kulkarni 2010). The most attractive reaction among those is the catalytic reforming of glycerol in aqueous or gaseous phase. When hydrogen is used, the fuel produces is clean and it can be used in a wide range of applications in many industries. Liquid hydrocarbons and alcohols (especially methanol) can be commercially converted from a mixture of CO and H_2 which is called as syngas as shown in Eq. 7.1 (James et al. 2010b). The similarity between steam reforming of glycerol and catalytic steam reforming of liquid hydrocarbons or natural gas is the use of supported metal catalysts (Holladay et al. 2009). Similarly, for the steam reforming of glycerol, the similar catalysts are often explored (Adhikari et al. 2009; Vaidya and Rodrigues 2009; Lin 2013). Metallic species such as Ru (Byrd et al. 2008), Pt (Kunkes et al. 2009), Rh (Chiodo et al. 2010), Ni (Iriondo et al. 2008; Cui et al. 2009), Ir (Zhang et al. 2007), and Co (Cheng et al. 2010a) have been confirmed to be the active catalysts for the steam reforming of glycerol:

$$(2n + 1)H_2 + nCO \rightarrow C_nH_{(2n+2)} + nH_2O \tag{7.1}$$

Generally, the steam reforming of glycerol is believed to involves at least two reactions in the catalytic reaction mechanism. Firstly, direct splitting of glycerol molecules is carried out to form CO and H_2 as shown in Eq. 7.2. And then. Additional H_2 can be yielded by water-gas shift reaction (WGS) when the steam (H_2O) is reacts with CO as shown in Eq. 7.3 (Adhikari et al. 2009; Chen and Syu 2010). Equation 7.4 shows the overall reaction of the processes (Vaidya and Rodrigues 2009).

$$C_3H_8O_3 \rightarrow 3CO + 4H_2 \qquad \Delta H = 251\,kJ/mol \tag{7.2}$$

$$H_2O + CO \rightarrow H_2 + CO_2 \qquad \Delta H = -41\,kJ/mol \tag{7.3}$$

$$C_3H_8O_3 + 3H_2O \rightarrow 7H_2 + 3CO_2 \qquad \Delta H = 128\,kJ/mol \tag{7.4}$$

As shown in Eq. 7.2, the splitting of glycerol reaction is highly endothermic. The heat taken by these splitting reaction is higher than the heat released in the next reaction involves WGS (Eq. 7.3) which is exothermic. Thus, the overall reaction of steam reforming of glycerol is endothermic (Chen and Syu 2010). With refer to the Eq. 7.4, in principle, the yield of hydrogen is favoured when low pressure, high ratios of water to glycerol and high temperatures is used (Adhikari et al. 2007). In addition, the target product of steam reforming of glycerol (either syngas or H_2) can be proceeded by adjusting the WGS reaction and also the splitting reaction of the glycerol through their reaction conditions and the catalysts used.

When supercritical conditions and separation and in situ adsorption of CO_2 were used, the significant improvement in the yield of hydrogen can be achieved as demonstrated in few recent studies. Clearly, the relatively large worldwide is in needed of the hydrogen to be used as clean energy carrier. If the production process of hydrogen by catalytic steam reforming of glycerol is commercialized, the productivity of glycerol in this present world would be regarded to be much limited. However, in the view of economy, such reactions on production of hydrogen remains controversial. Even though in hydrogen production for commercial scale, the steam reforming of hydrocarbons is primarily used in industry as the most efficient and economic technology. Firstly, the hydrogen contains in the each of glycerol molecule is less when compared to each molecule of hydrocarbon. Secondly, during the steam reforming of glycerol, much more sophisticated reactions can be occurred because glycerol is carbohydrate. As the result, cost purification of hydrogen will be higher because more compounds are generated simultaneously in the reactions. Therefore, reforming of glycerol in hydrogen production is currently uneconomic.

7.1.1 Supported Ni Catalysts

Primarily, due to its cheaper cost, Ni-base catalysts have been used for steam reforming of glycerol taken over the use of noble metal-based catalysts. Some of the examples which used Ni as their support includes, Ni/TiO_2, Ni/MgO, Ni/CeO_2 and Ni/Al_2O_3. Different catalytic performances can be observed between all those catalysts even though the same Ni species were used as their active components. For example, under same conditions where the temperature of reactions is 873 K, with 0.5 ml/min flow rate of the feed and molar ratio of 12:1 for glycerol to water, the selectivity to hydrogen were different for different catalysts. Where, the selectivity to hydrogen is 74.7, 38.6 and 28.3% for Ni/CeO_2, Ni/MgO and Ni/TiO_2 catalysts respectively (Adhikari et al. 2008). It proved that the choice of support did not effects their selectivity. The results suggested that the support screening and probing into the interactions between the active component and the support is important.

In addition, studies show that the catalytic performances in the steam reforming of the glycerol can be greatly improved if the promoters are properly doped to a Ni-based catalyst. Particularly, it have been proved that the use of species such as Zr, La and also Ce as the promoters will be a great help to Ni/Al_2O_3 catalysts (Dave and Pant 2011). On the one hand, the capability to activate reactants which are glycerol and steam can be increased when the promoters are present in a Ni-based catalyst (Iriondo et al. 2008). Besides that, the active Ni species also can be stabilize during the reactions if the proper promoters were used (Iriondo et al. 2008). For example, the stability of the Ni^0 particles were enhanced by the nickel-ceria interactions when $Ni/CeO_2/Al_2O_3$ catalysts were used (Iriondo et al. 2010). Besides, the functions to inhibiting the side reactions also can be provided when the appropriate support and promoter was used. For example, when $Ni-ZrO_2/CeO_2$ catalyst is used in the steam reforming of the glycerol, the ZrO_2 helps in a methanation reaction and the presence of CeO_2 prevented a dehydration reaction of glycerol. The selectivity to hydrogen in enhance by the former, and the formation of unsaturated hydrocarbons which is act as a precursor to the coke and lead to deactivation of catalyst is prevented by the latter. Few studies suggested that the size and also shape of the Ni-based catalyst is worth noting and very influential.

The shape of conventional Ni-based catalyst was either as spherical pellets or powders. Compared to those conventional shape, Ni-based catalyst with a monolithic structure was found to be more stable and can prevent the formation of coke (Bobadilla et al. 2016). Recently, it has been noticed the effect of metal ions when used in the aqueous solution of glycerol (Chen and Syu 2010). The presence of Na_2CO_3 in water was found to have negative effect on the yield of hydrogen under extreme conditions at temperature of 647 K and pressure of water is 22.1 MPa. On contrary, both hydrogen yield and glycerol conversion were enhanced with the presence of K_2CO_3 in water. One of the possible reason that the WGS reaction is promoted is the presence of the potassium ions (Chakinala et al. 2009). To achieve the steam reforming process which are economic feasible, these findings are very useful for future (Valliyappan et al. 2008b).

7.1.2   Supported Pt, Rh, and Ru Catalysts

When compared to the Ni-based catalysts, supported noble metal catalysts such as Rh, Pt and Ru are more stable and active in the steam reforming of glycerol processes (Gallo et al. 2010; Pompeo et al. 2010). Moreover, much better catalytic performances can be observed when bimetallic catalyst such as Pt–Re, Pt–Os and Pt–Ru were used. Generally, when supported noble metal catalysts were used in the reaction, lower temperatures which is less than 723 K are required. Recent studies on the effects of the supports acidity have provided a better understanding. Side reactions such as condensation and dehydration is catalysed by an acid support. For instance, because of the side reactions, formation of coke on the Pt/Al_2O_3 and Ru/La_2O_3 catalysts is easier. Even though the overall reforming reaction was enhanced, the tendency towards side reactions is increased when the temperatures of the reaction was increased. Therefore, the use of neutral support such as SiO_2 can solved this problem.

Alternatively, the negative effects of the acidic supports can be prevented by the use of short residence time and supercritical water, and it have been demonstrated by recent studies. As such, the use of short residence time and supercritical water had achieved a complete conversion of glycerol (May et al. 2010) with the use of Ru/ZrO_2 or Ru/Al_2O_3 (Byrd et al. 2008). Meanwhile, the increased in hydrogen yield was achieved and deposition of carbon was avoided. It has been proved that the reaction mechanism of steam reforming of glycerol is elusive and sophisticated. The reactions are highly depending on their reaction conditions and also the types of the catalysts used. Recently, it has been proven that, when WGS reaction is purposely intensified in the steam reforming of glycerol, the yield of hydrogen can be increased. A novel approach in which the WGS reaction was coupled with the steam reforming of glycerol have been demonstrated by Kunkes et al. (2009).

A reactor that is well-designed usually consist of two beds of catalysts. A bimetallic Pt–M/C that is use to convert glycerol into a mixture of CO/H_2 gases is included in the first bed. And the second bed consist of $Pt/CeO_2/ZrO_2$ catalyst which is used for the WGS reaction. Approximately 100% glycerol conversion with about 80% of hydrogen yield was achieved when such an integrated catalytic system is used, together with reaction conditions of temperature at 573 K under pressure of 0.1 MPa. Besides WGS reaction, the other reaction that can be coupled with the steam reforming of the glycerol with water is the exothermic partial oxidation of glycerol with air. To this end, the catalysts that are commonly used to catalyses both partial oxidation and steam reforming of glycerol is bifunctional supported noble metal catalysts.

Particularly, the use of energy in an effective way can be allowed when a thermoneutral process is establish. In this context, it must be understood that, for staying at the thermoneutral state their conditions are highly depends on the temperature, ratio of oxygen to glycerol and also ratio of steam to glycerol. Additionally, the formation of carbon and methane can be abated by the conditions of the thermoneutral. Nevertheless, the syngas production as the target product was favourable when the catalytic partial oxidation is combined with auto thermal steam reforming of glycerol as demonstrated by the catalytic performances using Rh–Ce as the catalysts (Rennard et al. 2009). The use of a membrane reactor that can acts simultaneously as a selective separator and as an active catalyst is expected to improve the yield of hydrogen. For instance, pure hydrogen was obtained when a steam reforming of glycerol is carried out by using a Pd–Ag membrane reactor as reported by Iulianelli et al. Moreover, higher conversion of glycerol can be achieved when membrane reactor was used instead of conventional reactor (Iulianelli et al. 2011).

Recently, the introduction of plasma catalytic and photocatalytic techniques to the steam reforming of the glycerol was carried out. For example, under much mild conditions, hydrogen can be yielded from the photocatalytic reforming of glycerol with the presence of catalysts such as Pd (Bowker et al. 2009), Au (Bowker et al. 2009) and Cu_2O/TiO_2 (Daskalaki and Kondarides 2009) or Pt/TiO_2 (Lalitha et al. 2010). However, it is still a hard task to scale up the plasma catalytic or photocatalytic processes.

Nowadays, increase in the severe climate change, high resource strain and also staggering environmental impact occurred as the results of enormous consumption of fossil fuels and also their derived chemical products. Therefore, increase in fractions of chemicals and fuels derived from renewable resources is expected by humans (Zhou et al. 2011; Gallezot 2012). Production of biodiesel through transesterification process of alcohol and triglycerides have been developed successfully. Nevertheless, demand for new uses of the glycerol which is the by-product was brought. Besides, abundance of glycerol can also be obtained from lignocellulosic biomass and also microorganisms. Therefore, in a new generation of renewable-based chemical industry, one of the compounds that can be a good platform is glycerol (Bozell and Petersen 2010; Zhou et al. 2008).

For the past five years, upsurge of interest in the catalytic conversion of glycerol to polymers, fuels, monomers and also fuel additives were indicated by scientific studies. When compared to pyrolysis, aqueous and gaseous phase reforming have captured more attention for the production of hydrogen from glycerol (Dou et al. 2009; Valliyappan et al. 2008a; Fernández et al. 2009). Information on factors such as pressure, contact time, temperature, ratio of water to glycerol and contaminants in feed that affects the yield of hydrogen have been provided in recent advances, however, not much information on the pilot test, reaction mechanism and also kinetics were reported. Coke formation stays as the main issues when Ni-based catalysts was used. It is still a challenge to carried out the process of glycerol reforming more economic. Alternatively, the coupled of glycerol reforming with other processes might solve the problems.

For instance, the heat produced from the synthesis of dimethyl ester was then used for the aqueous-phase reforming of glycerol which in return will produce the feedstock (syngas) that used in the synthesis of dimethyl ester. in that way, minimization of energy input can be done (Iliuta et al. 2012). Besides fuels, using glycerol in the production of additives for fuel is also worth further attention. Currently, there are several methods available, such as glycerol acetalization for production of ketals and acetals, glycerol etherification for the production of alkylglycerol and also glycerol esterification with acetic acid for the production of triacetin and diacetin. By using glycerol, alternative routes have been offered for production of valuable polymers and also monomers in large quantity (Parzuchowski et al. 2009; Ismail et al. 2010). Recently, some ground-breaking findings have been made.

For instance, when conversion of glycerol to 1,2-propanediol is carried out by catalytic hydrogenolysis, studies shows that excellent catalytic performances have been shown by Cu-based catalysts which is not expensive and its performance are comparable to the other catalysts which are expensive such as Ru-based, Pt-based and also Rh-based catalysts. Moreover, it was proved that, by excellent choice of catalysts, it is possible for the 1,3-propanediol to be the main product. Recently, studies show that conversion of glycerol to acrylonitrile and acrylic acid is also possible. In the production of epichlorohydrin, the glycerol chlorination to dichloropropanol has been carried out using an EpicerolTM process commercially. In conventional methods, the manufacturing of acrylonitrile, epichlorohydrin, 1,2-propanediol and 1,3-propanediol have been using petro-based as their feedstock. But now, more economical processes which are environment-friendly can also produced those compounds mentioned earlier by using glycerol-based processes. As such, production of these compounds using glycerol-based processes have been carried out by some companies such as Dow Chemical, DuPont and also Solvay Company. However, for commercialization purposes, there are some issues with the technological barriers involved. In most cases, the critical issues are to overcome the deactivation of catalysts used. Additionally, the crude glycerol that were produced from the biodiesel industry is expected to be used as the feedstock. Hence, the impurities in the crude glycerol must be endured by the catalyst. To solve these problem, when new reactors are developed, the choice of catalyst, product separation and also transport limitations must be considered. In this context, many choices were provided by the technology of process intensification. As shown in many recent studies, in the glycerol conversion, many methods can be used to intensify the catalytic reactions such as reactive distillation, membrane reactors, microwave heating and also supercritical fluids (Kongjao et al. 2011; Marshall and Haverkamp 2008).

However, for the catalytic conversion of glycerol, there are many other intensification methods that are still not explored such as oscillating flow, ultrasound, spinning disc and also rotating tube. These techniques are worth to pay attention to as integrated processes because they can also increase the rate of reaction and give benefits in safety, capital, environment protection and also energy (Xi and Davis 2010; Nascimento et al. 2011; Magnusson et al. 2010; Pan et al. 2011). For instance, potential in production of acrylonitrile, acrylic acid and lactic acid using the integrated dehydration-cyanation and oxidation-dehydration processes have been shown recently. Useful polymers that already have a large market can be produced directly from these products. In addition, it is noted that the glycerol can be used in many emerging methods.

By using glycerol, new functional which is biodegradable polymers, such as poly(glycerol-sebacate-L-lactide) copolymers (Liu et al. 2007) and poly(glycerol sebacate) elastome (Cheng et al. 2010b), can be made.

In addition, to produce 3-methylindole catalysed by Cu/SiO2-Al2O3 catalyst, reaction between glycerol and aniline can be done (Sun et al. 2010). It also has been proved that, when zeolite is used as catalyst, the direct conversion from glycerol to gasoline-range alkyl-aromatics is possible. Besides the catalytic conversion, it is also worth it to explore the other direct uses of glycerol. The glycerol was also being used directly as cosmetics, component in pharmaceuticals, food additives (Li et al. 2007), solvents (He et al. 2009) and also plastifiers (Mikkonen et al. 2009). Besides that, new processes can be created by integrated processes between biological conversion and chemical catalysis in order to use the glycerol efficiently (Anbarasan et al. 2012).

Production of fuels and chemicals such as amino acids, ethanol, n-butanol, polyols, 2,3-butanediol and succinic acid can be achieved through biological conversion of glycerol. By using chemocatalysis, these chemicals cannot be directly achieved from glycerol (Dobson et al. 2012; Ganesh et al. 2012; Khanna et al. 2012). Thus, by taking advantage of all these innovative approaches and theories, it is expected that the scientists and technologists will design good catalysts and reactors that have high performances and then new efficient processes with the use of glycerol can be established with less energy input and environmental harm.

7.2   Conversion of Biomass Derived Carbohydrates
          to Fuels

In recent years, the process on conversion of biomass-derived carbohydrates into furanic aldehydes have been the focused of many studies. This is because, these products can be further transformed into a wide range of high value-added chemicals and high performance fuels and they are considered as versatile intermediates. The example of furanic aldehydes includes 5-hydroxymethylfurfural (HMF), 5-halomethylfurfural, including 5-chloromethylfurfural (CMF) and 5-bromomethylfurfural (BMF), and also furfural. However, the production of furanic aldehydes were limited due to the special properties and chemical structure of the biomass-derived carbohydrates that lead to high production costs and also low yields. In order to overcome these limitations, development of various catalytic conversion routes have been carried out recently.

The worldwide attention has been attracted by the search for renewable resources due to growing concerns on global warming, diminishing fossil resources and also environmental pollution (Alonso et al. 2010; Geboers et al. 2011; Kobayashi et al. 2012; Rosatella et al. 2011; Tong et al. 2010; Zhou et al. 2011). With respect to this concern, biomass can be an ideal source that can be used to replace fossil resources due to their abundancy, low cost and also widespread (Binder and Raines 2010; Cao et al. 2011; Perego and Bosetti 2011; Qi et al. 2011; Wang et al. 2011; Serrano-Ruiz and Dumesic 2011). Through photosynthesis, about 170 billion metric tonnes of biomass have been produced by nature and 75% of it is assigned to carbohydrates (Lima et al. 2011; Ståhlberg et al. 2011; Zakrzewska et al. 2010; Qi et al. 2012). Therefore, for the utilisation of biomass-derived carbohydrates, the selection of efficient methods is very important. Nowadays, the conversion of biomass-derived carbohydrates into furanic aldehydes such as CMF, BMF, HMF and also furfural have been the most promising and attractive approaches which is used as a versatile platform to replace the fuels and chemicals which were derived from fossil resources as well as to synthesis wide range of new products (James et al. 2010a; Karinen et al. 2011; Liu et al. 2011; Serrano-Ruiz et al. 2011; Taarning et al. 2011).

Furfural is the product of xylose dehydration (Binder et al. 2010; Lam et al. 2012; Shi et al. 2011b; Zhang et al. 2012). It can be converted to linear alkanes (West et al. 2008), furfuryl alcohol, 2-methyltetrahydrofuran (MTHF) (Sitthisa and Resasco 2011; Serrano-Ruiz and Dumesic 2011), 2-methylfuran (MF) (Sitthisa and Resasco 2011), furoic acid, phenol-formaldehyde resin (Kim et al. 1994) and also maleic acid (Serrano-Ruiz et al. 2011; Shi et al. 2011a). The production of these products has been carried out in an industrial scale for decades. The production of furfural is concentrated in China with approximately 70% of global furfural production was produced there (Lima et al. 2011; Karinen et al. 2011; Mamman et al. 2008).

HMF is the product of hexose and it was first reported at the end of 19th century (Hansen et al. 2011; Hu et al. 2009; Qi et al. 2010; Román-Leshkov et al. 2006;

Vigier et al. 2012; Yong et al. 2008; Zhao et al. 2007). HMF consist of aromatic alcohol, aromatic aldehyde and also furan ring system, therefore, it is more multifunctional when compared to furfural (Tong et al. 2010; Lima et al. 2011; Zakrzewska et al. 2010). HMF can be converted to 2,5-dimethylfuran (DMF) (Román-Leshkov et al. 2006), 2,5-diformylfuran (DFF) (Xiang et al. 2011), 2,5-dihydroxymethylfuran (DHMF) (Román-Leshkov et al. 2006), levulinic acid (LA) (Peng et al. 2010), 2,5-dimethyltetrahydrofuran (DMTHF) (Yang and Sen 2010), linear alkanes (Huber et al. 2005) and also 2,5-furandicarboxylic acid (FDCA) (Gupta et al. 2011). More recently, two new platform chemicals of furanic aldehydes were reported by Kumari et al. (2011) and Mascal and Nikitin (2008) which are BMF and CMF, respectively. These BMF and CMF is similar to HMF and furfural in which they can also be converted into valuable fuels and chemicals, for example 5-ethoxymethylfurfural (EMF) and 2,5-dimethylfuran (DMF) (Mascal and Nikitin 2008).

However, due to their high production costs and low yields, the production of BMF, CMF and also HMF has never yet reached large scale and industrial scale. In the past few years, the production of BMF, CMF, HMF and also furfural from biomass-derived carbohydrates have been explored by some researches with the use of some low-cost and new high-efficiency processes. In 2008, the processes and technologies used for furfural production based on acid-catalysed hydrolysis of hemicellulose have been reviewed by Mamman et al. (2008). In 2011, the HMF formation through heterogeneous-catalysed and achievements in the ionic liquid-mediated have been summarized by Zakrzewska et al. (2010) and Karinen et al. (2011) respectively. However, there is no reports mentioned about the synthesis of BMF and also CMF.

The cost-competitive and high-effective processes on production of furanic aldehydes from the biomass-derived carbohydrates will decides for what industrial application the furanic aldehydes can be used. When compared to conventional catalytic conversion method, the production of furanic aldehydes used better solvents and catalysts and also cheaper materials (Lange et al. 2012). Homogeneous acid catalysts for example, H_3PO_4, CH_3COOH, H_2SO_4, HCOOH, HNO_3 and also HCl were used in the conventional method for furfural production (Mamman et al. 2008). However, these acid catalysts will give a higher environmental risks and they are also very corrosive (Serrano-Ruiz and Dumesic 2011; Lam et al. 2012). Recently, an environmental friendly and cleaner catalysts have been designed and proposed through several modifications of Lewis acids, solid acids and also various solvents. For instance, dehydration of xylose in water have been studied by (O'Neill et al. 2009) with the used of H-ZSM-5 catalyst, at the temperature of 200 °C for 18 min, and as the results, the yield of furfural is about 46%.

Besides that, the production of furfural from hemicellulose by using K10 and HUSY as the catalyst have been reported by Dhepe and Sahu (2010), gave 12% yields of furfural. In addition, compounds that can used in formation of furfural were synthesised, which are Sn-beta, MSHS-SO_3H, graphene, graphene oxide (GO), sulfonated graphene (SG) and sulfonated graphene oxide (SGO) and furfural yields of 14.3, 43.5, 51, 53.0, 55.0 and 62.0% were achieved, respectively

(Choudhary et al. 2011). It is worth it to note that for the synthesis of furfural, the most economical solvent that can be used is water. However, some undesired reactions can be accelerated and will results in the decrease of the furfural yields. Therefore, in the furfural production, the polar aprotic solvents were introduced.

About 37% yield of furfural from xylose can be achieved when the reaction was carried out in N,N-dimethylformamide (DMF) using Amberlyst-15 as the catalyst with addition of hydrotalcite (HT) as reported by Takagaki et al. (2010). While, about 60% yield of furfural from xylose can be obtained when the reaction was carried out in dimethyl sulfoxide (DMSO) using Nafion 117 as its catalyst. Even though to some extent the formation of the side-products can be avoided by using the polar aprotic solvents, they also have their own limitation in which their high boiling point and poor solubility of carbohydrates will results in problem to the separation and also production of the furfural (Lange et al. 2012).

In 1998, the presence of H-form mordenites and also faujasites in the mixed solvents of water and toluene were found to be effective for the production of furfural from xylose (Moreau et al. 1998). In biphasic systems, the dehydration process will take place in the aqueous phase and then as soon as it is formed, the furfural will be extracted to the organic phase (Moreau et al. 1998). By that mechanisms, the yields of the furfural can be improved and the unwanted secondary reactions can be reduced. Hence, many research groups further studied about the biphasic systems in which they are using various solid acids, such as sulphated metal oxides, zeolites, ion-exchange resins and also heteropoly acids with the yield of furfural obtained are ranging from 2.2 to 98%. Due to its less corrosiveness, easy to be separated and easy to be reused, solid acids are an attractive option to be used in the synthesis of furfural. However, after a few cycles of reaction, loss of active sites and humins deposition might occurred and results in the deactivation of the solid acids. In that case, frequent regeneration of solid acids need to be done by impregnation and calcination processes (Lange et al. 2012).

In recent years, ionic liquids have been successfully used in the furfural production from xylan and xylose. It can be due to its specific properties such as non-flammability, low melting point, high thermal stability, remarkable solubilizing ability, close to infinite structural variation and negligible vapor pressure (Lima et al. 2011; Zakrzewska et al. 2010). An effective production of furfural from xylose have been reported by Lima et al. (2011) and Tao et al. (2010) using acidic ionic liquids 1-ethyl-3-methylimidazolium hydrogen sulfate ([EMIM][HSO$_4$]) and 1-(4-sulfonic acid)butyl-3-methylimidazolium hydrogen sulfate ([SBMIM][HSO$_4$]) as both solvents and catalysts with the furfural yields of 84 and 91.5%, respectively. When H$_2$SO$_4$ was used as catalyst, the use of 1-butyl-3-methylimidazolium chloride ([BMIM]Cl) which is a neutral ionic liquid as solvent also lead to an active reaction.

The conversion of xylose to furfural in the presence of N,N-dimethylacetamide (DMA) have been investigated by Binder et al. (2010) using CrCl$_2$ as catalyst. From the results, the improvement in the furfural yield can be achieved when 1-butyl-3-methylimidazolium bromide ([BMIM]Br) is added. About 25% of furfural yield was achieved when xylan was used as the substrate in [EMIM]Cl by the use of

CrCl$_2$ and HCl. In addition, the yield of furfural can be up to 63% when the xylan is converted in [BMIM]Cl under microwave irradiation with addition of CrCl$_3$ as demonstrated by Zhang and Zhao (2010). Even though the results obtained were good, for the long-term production and recyclability, the use of ionic liquids is very critical because they are very expensive. Besides that, there is no methods on the purification and separation of the furfural from the ionic liquid have been fully demonstrated.

Previous studies reported that there are two possible pathways involved in the production of furfural from xylose (Ahmad et al. 1995; Feather 1970; Marcotullio and De Jong 2010). The first one is based on cyclic intermediates (Nimlos et al. 2006) and the second one is based on acyclic intermediates. In the formation of furfural, the process of hydrogen transfer in the isomerization of xylose to xylulose is thought to be the rate-limiting step because it is much slower than the dehydration of xylulose (Binder et al. 2010; Choudhary et al. 2011). The hydrogen transfer can occur either by 1,2-hydride shift mechanism or by the 1,2-enediol mechanism. However, the debates on the mechanism for formation of furfural is still going on and it must be further investigated. The mechanism is believed to be depends on the solvents, catalyst and also the conditions of the reactions used.

Over the past decades, concerns in both industrial and scientific communities have been attracted to the catalytic conversion of biomass-derived carbohydrates through furanic aldehydes to produce chemicals and also fuels. Until now, there is no large-scale production of furanic aldehydes have been carried out even though they have shown some exciting results in laboratory scale. Some point should be addressed in the future studies in order to have large-scale production of furanic aldehyde from biomass-derived carbohydrates. Firstly, the kinetics and also mechanisms of the dehydration reaction of biomass-derived carbohydrates must be further investigated. And it can be done by using the joint method of density functional theory, molecular dynamics, nuclear magnetic resonance, energetics and also quantum mechanics.

Besides that, the catalysts used must be well-designed to be more stable, active and also green catalysts. The design and preparation of the ionic liquid also must be strategic to have ionic liquid that is biodegradable, less corrosive, less viscous and the most important thing is the ionic liquid must be cheaper. Next, for the production of furanic aldehydes to be more convenient and faster, an effective combination of ionic liquids, microwave irradiation and solid catalysts should be prepared and designed. The cost and separation steps of furanic aldehydes can be reduced with single step synthesis where the biomass-derived carbohydrates can be directly converted to furanic aldehydes-based fuels and chemicals. Therefore, establishment of single step conversion will be highly appreciated.

Establishment of technologies that is energy-efficient and have high efficiency such as supercritical carbon dioxide extraction and membrane separation also will be very useful in purification and separation steps to get the desired products. Then, to evaluate the overall processes on the furanic aldehydes production, a complete techno-economic evaluation should be done. Besides that, the assessment on the products (furanic aldehydes and related fuels and chemicals) should be carried out to make sure that the products are not harmful to both environment and also humans. Lastly, the development of technologies for continuous production and optimisation of the reaction conditions must be designed and investigated, respectively. All in all, the efforts in looking for environment friendly, cost-effective and also workable methods to be used in large-scale productions of chemicals and fuels from biomass-derived carbohydrates through furanic aldehydes must be continued until it was achieved.

7.3 Catalysis in Depolymerisation of Lignin

Throughout the modern society, the concept of sustainability is embedded. This concept can be seen from the methods of transport and food production and also design of the cities, to the recycling and sourcing of the natural resources. Together with these development, new technologies have been able to deliver fuels and also chemicals from the renewable resources, in which biomass has become one of the most common renewable resources used as feedstocks due to its abundancy and also economically attractive.

Particularly, due to its low carbon energy, lignocellulosic biomass is considered to be one of the most attractive source of biomass, ant the only limitation faces by these source is the competition from the paper and agricultural sector. However, due to the presence of halogen elements and water-soluble fraction of alkaline, high oxygen and water content, and also hazardous trace element, which will result in low-energy heating/density value, ash-fusion temperatures and pH of biomass, it leads to the problematic issues of using biomass as a replacement for fossil-based feedstocks. Further development of versatile method on chemical transformation of biomass is needed due to their heterogeneous nature of biomass between different microorganisms and plants, and also within the same species at different regions (Vassilev et al. 2015).

However, there are strong financial and political drivers for technologies of renewable energy with mandates from European Union where by 2020, 20% of overall energy consumption must be derived from renewable sources and increasing to 27% by 2030. Besides that, 10% of the transportation fuels must be derived from renewable resources, and when compared to fossil-based fuels, at least 35% reductions in greenhouse gases emissions must be achieved. In order to make sure that the land which was designed for bio-fuels production from biomass was previously not used for carbon stock, for example forest or wetlands, or impact regions that have high biodiversity, for example primary forests, policy directives was also introduced (Union 2009).

Lignocellulosic biomass composed of cellulose, lignin, and also hemicellulose, with lignin as the second most abundant constituent in the lignocellulosic after cellulose at around 10 to 30 wt%. lignin consist of highly branched phenolic polymer with high molecular weight, usually in between 600 and 1500 kDa (Kleinert and Barth 2008) with around 40–50 million tonnes of annual global production. Lignin usually forms in the cell walls of plants and its composition consist of syringyl alcohol, p-coumaryl alcohol and also guaiacyl alcohol (White and Kennedy 1985). Depolymerisation of lignin can afford the 'platform chemicals' of the phenolic, and subsequently much attention has been attracted in order to be used in the production of both bio-fuels and also chemicals. In between the monomers of lignin, there are C–O bonds. And it is quite challenging for the selective cleavage of that C–O bonds to take place. However, the compounds produce is suitable for the upgrading process for applications which involves fine chemicals, and the compounds can be achieved via hydrolysis, solvolysis, pyrolysis, alkaline oxidation and also hydrogenolysis.

Due to radical formation, usually the oxidizing routes is not desirable. This is because, radical formation can lead to partial repolymerization, meanwhile the protocols of hydrogenolysis is to promote the bond cleavage of C–O and minimizing the lignin repolymerization, through recombination of radicals and also quenching, even though fully hydrogenated cyclic hydrocarbons with low commercial value can be achieved when the molecular hydrogen is added. It is promising that the hydrogen can be generated or transferred from the hydrogen-donating solvents. For example, during hydrocracking of lignin, hydrogen transfer from formic acid, ethanol and also tetralin have been reported by Connors et al. (1980). The CO_2 and hydrogen was liberated from a solvolysis method that used high-pressure thermal treatment of lignin, in the presence of formic acid as the hydrogen donor and also ethanol as solvent was reported by Kleinert and Barth (2008).

The other promising hydrogen-donor solvents are glycerol (Wolfson et al. 2009) and isopropanol (Johnstone et al. 1985). In order to selectively deoxygenate the generated products and to facilitate the atom and energy-efficient lignin depolymerisation, the use of catalysts is very important. This is due to the strength of the C–O bonds that held the lignin network together and also due to the various possible products that can be produced. Through the biorefinery concept, the key for production of chemicals and fuels from biomass is the lignin valorisation. Three broad classes of potential process under consideration within the biorefineries have been categorised by Gallezot (Kim et al. 2014). First, to deliver condensable or gaseous molecules for the next transformation through established chemical processing, high-temperature thermal conversion by pyrolysis or gasification must be used. Second, lower temperature enzymatic or catalytic conditions must be used to convert lignin into aromatic building blocks. And third, highly functionalised products can be generated directly through one-pot routes.

Due to its complex interconnectivity among the polymeric structure of the lignin, the decomposition of lignin to phenolic components is hindered (Sturgeon et al. 2014). For production of phenolic compounds from lignin, several methods including oxidation, hydrolysis and hydrogenolysis (hydrocracking) have been carried out (Pandey and Kim 2011). Oxidative cleavage of C–O–C and C–C bonds can result in vanillin and similar compounds together with formation of CO, CO_2, and H_2O. However, due to the production of free radicals as the side product that will results in re-condensation, the oxidative protocols become undesirable. In contrast, it has been proved that the reductive cleavage of such bonds is more successful in yielding the monomeric compounds. Due to their disadvantages with respect to their corrosiveness, difficulty in separation from reaction mixture, and, favoured for selective bond cleavage in the model compounds, but have been implemented in oxidative depolymerisation, the utilisation of the homogeneous catalysts were rarely seen (Behling et al. 2016).

Excellent route to produce monomeric phenols can be seen when lignin depolymerisation was carried out together with base catalysts. In base-catalysed depolymerisation of lignin, it is much easier to cleave aryl–alkyl (β-O-4) ether bonds when compared to C–C bonds (Roberts et al. 2010), even though extremely high reaction temperatures of around 340 °C is usually used for mineral bases (Thring 1994), where it can produced significant gaseous products through side reactions and consequent lower monomeric hydrocarbon yields (Xu et al. 2014). The choice of lignin used in the base-catalysed depolymerisation will only effects the bio-oil yield, but it does not effects the composition of the resulting products (Charlier et al. 1991).

However, neutralisation of the bio-oil is needed when mineral bases is used, with addition to the corrosion of reactor. In the production of syringol, guaiacol and catechol with its derivatives through the depolymerisation of lignin of Alcell, the effect of temperature and time was studied by Thring et al. in the presence of NaOH as the catalyst (Thring 1994; Roberts et al. 2011). Effective methods on the base-catalysed depolymerisation in polar organic solvents, such as methanol and ethanol, have been proven by enhancing the production of dimer and monomer relative to solid acid catalysts and also suppressing the formation of char. The repolymerisation of monomeric products from lignin can be prevented by using base catalysts, and as reported by Toledano et al. (2014), the prevention from repolymerisation will results in higher yields of bio-oils (Erdocia et al. 2014). During mechanism of depolymerisation, the strength of base will plays an important role, where the production of cresol, phenol, 4-methylcatecol and catechol were promoted by hydroxides, while when potassium carbonate was used, neither cresols or catechol were observed (Erdocia et al. 2014).

In order to minimise repolymerisation and obtained products with lower molecular weight, capping agents such as boric acid have been used in order to protect the hydroxyl groups in the phenolic compounds (Roberts et al. 2011). In alkaline media, production of respective esters ($NaB(OR)_4$) by reaction of phenolic monomers and boric acid will prevents the side reactions of the phenol products, and at the same time formation of char also can be prevented. When compared to the single protection using boric acid under analogous conditions, higher yields can also be offered by borated esters formation during base-catalysed depolymerisation. The monomeric products of the depolymerisation of lignin can be stabilise by using phenol as the capping agent, followed by formation of ferulic acid and catechol, while, significant dimers can be produced when boric acid was used as the capping agent. The ideas on different behaviour between these two capping agents have been discussed by Toledano et al. (2014). For boric acid, the monomeric aromatics yields are lowered because of the intermediates trapping by the boric acid. Whereas, oligomerisation was reduced by capping of phenol, but not dealkylation and also demethoxylation pathways.

In the synthesis of biodiesel, MgO as the solid base catalysts has been used widely (Montero et al. 2009). Besides that, it has been used in depolymerisation of lignin in a range of solvents such as ethanol, water, tetrahydrofuran and also methanol. When tetrahydrofuran was used, maximum conversion of lignin can be achieved, and it can be due to the superior solubilisation of lignin in the tetrahydrofuran. However, the repolymerisation of phenol oligomers also can be catalysed by MgO.

A promising cleavage of β-O-4 of ether bonds in the lignin compounds were also shown with the other solid bases such as $RbCO_3$ $CsCO_3$. In addition, the yields of the bio-oil can be increased and the formation of the char also can be suppressed (Dabral et al. 2015). The combination of these active compounds (heterogeneous Ni catalysts and homogeneous base catalysts) have been proposed by Sturgeon et al. (2014), this is because, from previous discussion, it have been proved that high conversion yields (95%) (Roberts et al. 2011) and ability to cleave the C–O bonds (Sergeev et al. 2012) have been shown by these homogeneous base catalysts and heterogeneous Ni catalysts respectively.

Depolymerisation of lignin model compound, 2-phenoxy-1-phenethanol, extraction of ball-milled lignin from corn stove into alkyl aromatics that have low molecular weight have been shown in a reaction that was catalysed by a 5 wt% Ni supported on a MgAl hydrotalcite (Woodford et al. 2012; Creasey et al. 2015). Interestingly, in order to be used in the catalytic application, there is no activation with H_2 is needed for the nickel nanoparticles that present on the external surface of the hydrotalcite, and it suggested that the active species was a mixed of nickel oxide valence. The interlayer anion nature is strongly influence the catalytic properties of the nickel hydrotalcite (Kruger et al. 2016). Even though it is still not clear about the origin of the base promotion, direct nitration of the phenolic might be involved in the mechanism of the depolymerisation. The depolymerisation of lignin from corn stove which is enzymatically hydrolysed using the same nitrate intercalated nickel hydrotalcite was also proved to be effective. When the process is performed in water at the temperature of less than 200 °C, large amounts of products with low molecular weight such as syringol, guaiacol and also phenol monomers can be produced.

Due to their propensity to generates free radicals that will lead to production of char and also repolymerization of mono or dimeric products, the oxidative depolymerisation is considered as undesirable. However, because the retention of both acyclic and aromatic organic frameworks without the cleavage of C–C bonds can be permitted using this process, oxidative reactions can also be advantageous. Oxidative degradation have been adopted as a route for decomposition of lignin (Zhao et al. 2013), and generates vanillin through a hydrogen-free method (Ragauskas et al. 2014). For the oxidative depolymerisation of lignin, a range of organometallics catalysts such as vanadium and polyoxometalates have been reported. For the cleavage of C–O bonds in the compounds of lignin model, complexes of vanadium-oxo have been demonstrated to be an efficient catalysts (Son and Toste 2010).

As reported by Silk et al., the selectivity that will directing either the scission of C–O linkages or cleavage of C–C bond was shown to be strongly influenced by the choice of the vanadium complex (Korpi et al. 2006). Polyoxometalates is an effective catalyst to be used in the oxidative degradation and it is also a robust redox catalyst. Process which involves two-step will occurs if the polyoxometalate catalyst is performed in the presence of oxygen, first reduction of polyoxoanion and the second step it will be reoxidised with no changes in the structure (Ammam 2013). H_2O and CO_2 is commonly yielded as the by-products when the polyoxometalates catalyst was applied in the decomposition reaction of lignin residues under oxygen (Gaspar et al. 2007). Under oxygen, when pyrolytic of lignin was carried out using molybdenum or vanadium polyoxometalate catalysts in a mixture of methanol/water solvents, the aromatics and dimethyl succinate will also be produced (Zhao et al. 2013).

When the amount of methanol via methanol oxidation is increased, the production of bio-oils also can be increased, this is because the depolymerisation of the lignin was promoted by the CH_3OC and CCH_3 radicals that were produced by oxidation. However, the production of monomeric products is not favoured under alkaline conditions even though addition of methanol is carried out (Voitl and Rudolf von Rohr 2008). Through the aromatics breakdown, major products consist of esters and organic acids such as dimethyl succinate and dimethyl fumarate were formed when $H_5PMo_{10}V_2O_{40}$ is used as the catalyst (Kamwilaisak and Wright 2012). By using either molecular oxygen or hydrogen peroxide, the transformation of oxidative lignin also have been tested by using Salen complexes, especially cobalt as the catalyst. The oxidative transformation of the lignin compounds also can be catalysed by copper complexes catalyst for example Cu-phen and also Cu-bipy. Oxidation activity is dependent on the of electronic ligand and steric effects.

Transformation of Organosolv lignin in ethanol with the used of bifunctional catalyst under oxygen have been successfully produced guaiacol, vanillin and also 4-hydroxybenzaldehyde. By using molecular oxygen as the oxidant, oxidative cleavage of dilignol model compound have been carried out with the transition metal containing hydrotalcite as the catalyst. The selection of solvent in the oxidative depolymerisation will influenced the yield conversion, where, negligible conversion can be observed when toluene is used as the solvent, while in pyridine good activity obtained with the veratric acid is produced as the major product. The same catalysts were then applied to both Kraft and Organosolv lignin. However, due to the complexity of zeolite contaminants presences in the source and also practical lignin sources, their performances could not be extrapolated from the dilignol model. Therefore, more extreme conditions are required in order to depolymerise the Organosolv lignin.

For production of bio-fuels and also chemicals, the importance of renewable resources is representing by lignin. Therefore, its valorisation will play and important role in the sustainable biorefineries development. However, the main limitation to established one standard methods for depolymerisation of lignin is their heterogeneous nature. When a novel catalytic route for depolymerisation of lignin have been developed, it will result in energy-efficient and atom-efficient processes, and also safer approaches.

For the decomposition of lignin to carried out at low temperature, alternative solvents for example ILs can be used in conjunction with the conventional mineral acids or tailored heterogeneous or homogeneous catalysts, even though the potential corrosiveness, high price and also toxicity of some of the ILs will limits their applications in industrial scale.

The aromatic fraction that was generated from catalytic lignin depolymerisation still have a comparatively high acid content and low heating value, as the results of the high proportion of oxygenated compounds present in the bio-oil, and concomitant poor miscibility with aliphatic hydrocarbons, in addition to a broad boiling point range. Therefore, it is very important for both chemical and fuel applications to have the technologies in bio-oil upgrading. Hydrodeoxygenation (HDO) reaction was proved to be the most successful technology used for biofuel, in which, complete hydrogenation of oxygenated depolymerisation products to benzene and cyclohexane can be obtained.

References

Adhikari S, Fernando S, Gwaltney SR, To SF, Bricka RM, Steele PH, Haryanto A (2007) A thermodynamic analysis of hydrogen production by steam reforming of glycerol. Int J Hydrogen Energy 32(14):2875–2880

Adhikari S, Fernando SD, To SF, Bricka RM, Steele PH, Haryanto A (2008) Conversion of glycerol to hydrogen via a steam reforming process over nickel catalysts. Energy Fuels 22 (2):1220–1226

Adhikari S, Fernando SD, Haryanto A (2009) Hydrogen production from glycerol: an update. Energy Convers Manag 50(10):2600–2604

Ahmad T, Kenne L, Olsson K, Theander O (1995) The formation of 2-furaldehyde and formic acid from pentoses in slightly acidic deuterium oxide studied by ^{1}H NMR spectroscopy. Carbohyd Res 276(2):309–320

Alonso DM, Bond JQ, Dumesic JA (2010) Catalytic conversion of biomass to biofuels. Green Chem 12(9):1493–1513

Ammam M (2013) Polyoxometalates: formation, structures, principal properties, main deposition methods and application in sensing. J Mater Chem A 1(21):6291–6312

Anbarasan P, Baer ZC, Sreekumar S, Gross E, Binder JB, Blanch HW, Clark DS, Toste FD (2012) Integration of chemical catalysis with extractive fermentation to produce fuels. Nature 491 (7423):235–239

Arroyo-López FN, Pérez-Torrado R, Querol A, Barrio E (2010) Modulation of the glycerol and ethanol syntheses in the yeast Saccharomyces kudriavzevii differs from that exhibited by Saccharomyces cerevisiae and their hybrid. Food Microbiol 27(5):628–637

Atadashi I, Aroua M, Aziz AA, Sulaiman N (2012) The effects of water on biodiesel production and refining technologies: a review. Renew Sustain Energy Rev 16(5):3456–3470

Ayoub M, Abdullah AZ (2012) Critical review on the current scenario and significance of crude glycerol resulting from biodiesel industry towards more sustainable renewable energy industry. Renew Sustain Energy Rev 16(5):2671–2686

Behling R, Valange S, Chatel G (2016) Heterogeneous catalytic oxidation for lignin valorization into valuable chemicals: what results? What limitations? What trends? Green Chem 18 (7):1839–1854

Binder JB, Raines RT (2010) Fermentable sugars by chemical hydrolysis of biomass. Proc Natl Acad Sci 107(10):4516–4521

Binder JB, Blank JJ, Cefali AV, Raines RT (2010) Synthesis of furfural from xylose and xylan. Chemsuschem 3(11):1268–1272

Bobadilla L, Blay V, Álvarez A, Domínguez M, Romero-Sarria F, Centeno M, Odriozola J (2016) Intensifying glycerol steam reforming on a monolith catalyst: a reaction kinetic model. Chem Eng J 306:933–941

Bournay L, Casanave D, Delfort B, Hillion G, Chodorge J (2005) New heterogeneous process for biodiesel production: a way to improve the quality and the value of the crude glycerin produced by biodiesel plants. Catal Today 106(1):190–192

Bowker M, Davies PR, Al-Mazroai LS (2009) Photocatalytic reforming of glycerol over gold and palladium as an alternative fuel source. Catal Lett 128(3–4):253

Bozell JJ, Petersen GR (2010) Technology development for the production of biobased products from biorefinery carbohydrates—the US Department of Energy's "Top 10" revisited. Green Chem 12(4):539–554

Brandner A, Lehnert K, Bienholz A, Lucas M, Claus P (2009) Production of biomass-derived chemicals and energy: chemocatalytic conversions of glycerol. Top Catal 52(3):278–287

Byrd AJ, Pant K, Gupta RB (2008) Hydrogen production from glycerol by reforming in supercritical water over Ru/Al$_2$O$_3$ catalyst. Fuel 87(13):2956–2960

Cao Q, Guo X, Guan J, Mu X, Zhang D (2011) A process for efficient conversion of fructose into 5-hydroxymethylfurfural in ammonium salts. Appl Catal A 403(1):98–103

Carlson TR, Tompsett GA, Conner WC, Huber GW (2009) Aromatic production from catalytic fast pyrolysis of biomass-derived feedstocks. Top Catal 52(3):241

Carmona M, Valverde JL, Pérez A, Warchol J, Rodriguez JF (2009) Purification of glycerol/water solutions from biodiesel synthesis by ion exchange: sodium removal Part I. J Chem Technol Biotechnol 84(5):738–744

Chakinala AG, Brilman DW, van Swaaij WP, Kersten SR (2009) Catalytic and non-catalytic supercritical water gasification of microalgae and glycerol. Ind Eng Chem Res 49(3):1113–1122

Charlier J-C, Michenaud J-P, Gonze X, Vigneron J-P (1991) Tight-binding model for the electronic properties of simple hexagonal graphite. Phys Rev B 44(24):13237

Chen W-H, Syu Y-J (2010) Hydrogen production from water gas shift reaction in a high gravity (Higee) environment using a rotating packed bed. Int J Hydrogen Energy 35(19):10179–10189

Cheng CK, Foo SY, Adesina AA (2010a) H$_2$-rich synthesis gas production over Co/Al$_2$O$_3$ catalyst via glycerol steam reforming. Catal Commun 12(4):292–298

Cheng S, Yang L, Gong F (2010b) Novel branched poly (l-lactide) with poly (glycerol-co-sebacate) core. Polym Bull 65(7):643–655

Chiodo V, Freni S, Galvagno A, Mondello N, Frusteri F (2010) Catalytic features of Rh and Ni supported catalysts in the steam reforming of glycerol to produce hydrogen. Appl Catal A 381 (1):1–7

Choudhary V, Pinar AB, Sandler SI, Vlachos DG, Lobo RF (2011) Xylose isomerization to xylulose and its dehydration to furfural in aqueous media. Acs Catalysis 1(12):1724–1728

Connors W, Johanson L, Sarkanen K, Winslow P (1980) Thermal degradation of kraft lignin in tetralin. Holzforschung-Int J Biol Chem Phy Technol Wood 34(1):29–37

Creasey JJ, Parlett CM, Manayil JC, Isaacs MA, Wilson K, Lee AF (2015) Facile route to conformal hydrotalcite coatings over complex architectures: a hierarchically ordered nanoporous base catalyst for FAME production. Green Chem 17(4):2398–2405

Cui Y, Galvita V, Rihko-Struckmann L, Lorenz H, Sundmacher K (2009) Steam reforming of glycerol: the experimental activity of La1–xCexNiO$_3$ catalyst in comparison to the thermodynamic reaction equilibrium. Appl Catal B 90(1):29–37

Cydzik-Kwiatkowska A, Wojnowska-Baryła I, Selewska K (2010) Granulation of sludge under different loads of a glycerol fraction from biodiesel production. Eur J Lipid Sci Technol 112 (5):609–613

Dabral S, Mottweiler J, Rinesch T, Bolm C (2015) Base-catalysed cleavage of lignin β-O-4 model compounds in dimethyl carbonate. Green Chem 17(11):4908–4912

Daskalaki VM, Kondarides DI (2009) Efficient production of hydrogen by photo-induced reforming of glycerol at ambient conditions. Catal Today 144(1):75–80

Dave CD, Pant K (2011) Renewable hydrogen generation by steam reforming of glycerol over zirconia promoted ceria supported catalyst. Renew Energy 36(11):3195–3202

Devi P, Bethala L, Gangadhar KN, Sai Prasad PS, Jagannadh B, Prasad RB (2009) A Glycerol-based carbon catalyst for the preparation of biodiesel. Chemsuschem 2(7):617–620

Dhepe PL, Sahu R (2010) A solid-acid-based process for the conversion of hemicellulose. Green Chem 12(12):2153–2156

Dobson R, Gray V, Rumbold K (2012) Microbial utilization of crude glycerol for the production of value-added products. J Ind Microbiol Biotechnol 39(2):217–226

Dou B, Dupont V, Williams PT, Chen H, Ding Y (2009) Thermogravimetric kinetics of crude glycerol. Biores Technol 100(9):2613–2620

Erdocia X, Prado R, Corcuera MA, Labidi J (2014) Base catalyzed depolymerization of lignin: influence of organosolv lignin nature. Biomass Bioenergy 66:379–386

Fan Y, Zhou C, Zhu X (2009) Selective catalysis of lactic acid to produce commodity chemicals. Catal Rev 51(3):293–324

Feather MS (1970) The conversion of D-xylose and D-glucuronic acid to 2-furaldehyde. Tetrahedron Lett 11(48):4143–4145

Fernández Y, Arenillas A, Díez MA, Pis J, Menéndez J (2009) Pyrolysis of glycerol over activated carbons for syngas production. J Anal Appl Pyrol 84(2):145–150

Fernández Y, Arenillas A, Bermúdez JM, Menéndez J (2010) Comparative study of conventional and microwave-assisted pyrolysis, steam and dry reforming of glycerol for syngas production, using a carbonaceous catalyst. J Anal Appl Pyrol 88(2):155–159

Gallezot P (2012) Conversion of biomass to selected chemical products. Chem Soc Rev 41 (4):1538–1558

Gallo A, Pirovano C, Marelli M, Psaro R, Dal Santo V (2010) Hydrogen production by glycerol steam reforming with Ru-based catalysts: a study on Sn doping. Chem Vap Deposition 16(10–12):305–310

Ganesh I, Ravikumar S, Hong SH (2012) Metabolically engineered Escherichia coli as a tool for the production of bioenergy and biochemicals from glycerol. Biotechnol Bioprocess Eng 17 (4):671–678

Gaspar AR, Gamelas JA, Evtuguin DV, Neto CP (2007) Alternatives for lignocellulosic pulp delignification using polyoxometalates and oxygen: a review. Green Chem 9(7):717–730

Geboers JA, Van de Vyver S, Ooms R, de Beeck BO, Jacobs PA, Sels BF (2011) Chemocatalytic conversion of cellulose: opportunities, advances and pitfalls. Catal Sci Technol 1(5):714–726

Gupta NK, Nishimura S, Takagaki A, Ebitani K (2011) Hydrotalcite-supported gold-nanoparticle-catalyzed highly efficient base-free aqueous oxidation of 5-hydroxymethylfurfural into 2,5-furandicarboxylic acid under atmospheric oxygen pressure. Green Chem 13(4):824–827

Hadi M, Shariati M, Afsharzadeh S (2008) Microalgal biotechnology: carotenoid and glycerol production by the green algae Dunaliella isolated from the Gave-Khooni salt marsh. Iran Biotechnol Bioprocess Eng 13(5):540–544

Hájek M, Skopal F (2010) Treatment of glycerol phase formed by biodiesel production. Biores Technol 101(9):3242–3245

Hansen TS, Mielby J, Riisager A (2011) Synergy of boric acid and added salts in the catalytic dehydration of hexoses to 5-hydroxymethylfurfural in water. Green Chem 13(1):109–114

He F, Li P, Gu Y, Li G (2009) Glycerol as a promoting medium for electrophilic activation of aldehydes: catalyst-free synthesis of di(indolyl) methanes, xanthene-1,8(2H)-diones and 1-oxo-hexahydroxanthenes. Green Chem 11(11):1767–1773

Holladay JD, Hu J, King DL, Wang Y (2009) An overview of hydrogen production technologies. Catal Today 139(4):244–260

Hu S, Zhang Z, Song J, Zhou Y, Han B (2009) Efficient conversion of glucose into 5-hydroxymethylfurfural catalyzed by a common Lewis acid $SnCl_4$ in an ionic liquid. Green Chem 11(11):1746–1749

Huber GW, Chheda JN, Barrett CJ, Dumesic JA (2005) Production of liquid alkanes by aqueous-phase processing of biomass-derived carbohydrates. Science 308(5727):1446–1450

Iliuta I, Iliuta MC, Fongarland P, Larachi F (2012) Integrated aqueous-phase glycerol reforming to dimethyl ether synthesis—a novel allothermal dual bed membrane reactor concept. Chem Eng J 187:311–327

Ilyushin V, Motiyenko R, Lovas F, Plusquellic D (2008) Microwave spectrum of glycerol: observation of a tunneling chiral isomer. J Mol Spectrosc 251(1):129–137

Iriondo A, Barrio V, Cambra J, Arias P, Güemez M, Navarro R, Sánchez-Sánchez M, Fierro J (2008) Hydrogen production from glycerol over nickel catalysts supported on Al_2O_3 modified by Mg, Zr, Ce or La. Top Catal 49(1–2):46–58

Iriondo A, Barrio V, Cambra J, Arias P, Guemez M, Sanchez-Sanchez M, Navarro R, Fierro J (2010) Glycerol steam reforming over Ni catalysts supported on ceria and ceria-promoted alumina. Int J Hydrogen Energy 35(20):11622–11633

Ismail TNMT, Hassan HA, Hirose S, Taguchi Y, Hatakeyama T, Hatakeyama H (2010) Synthesis and thermal properties of ester-type crosslinked epoxy resins derived from lignosulfonate and glycerol. Polym Int 59(2):181–186

Israel A, Obot I, Asuquo J (2008) Recovery of glycerol from spent soap LyeBy-product of soap manufacture. J Chem 5(4):940–945

Iulianelli A, Seelam P, Liguori S, Longo T, Keiski R, Calabro V, Basile A (2011) Hydrogen production for PEM fuel cell by gas phase reforming of glycerol as byproduct of bio-diesel. The use of a Pd–Ag membrane reactor at middle reaction temperature. Int J Hydrogen Energy 36(6):3827–3834

James OO, Maity S, Usman LA, Ajanaku KO, Ajani OO, Siyanbola TO, Sahu S, Chaubey R (2010a) Towards the conversion of carbohydrate biomass feedstocks to biofuels via hydroxylmethylfurfural. Energy Environ Sci 3(12):1833–1850

James OO, Mesubi AM, Ako TC, Maity S (2010b) Increasing carbon utilization in Fischer-Tropsch synthesis using H_2-deficient or CO_2-rich syngas feeds. Fuel Process Technol 91(2):136–144

Jerzykiewicz M, Cwielag I, Jerzykiewicz W (2009) The antioxidant and anticorrosive properties of crude glycerol fraction from biodiesel production. J Chem Technol Biotechnol 84(8):1196–1201

Johnson DT, Taconi KA (2007) The glycerin glut: options for the value-added conversion of crude glycerol resulting from biodiesel production. Environ Prog 26(4):338–348

Johnstone RA, Wilby AH, Entwistle ID (1985) Heterogeneous catalytic transfer hydrogenation and its relation to other methods for reduction of organic compounds. Chem Rev 85(2):129–170

Kaçka A, Dönmez G (2008) Isolation of Dunaliella spp. from a hypersaline lake and their ability to accumulate glycerol. Biores Technol 99(17):8348–8352

Kale GR, Kulkarni BD (2010) Thermodynamic analysis of dry autothermal reforming of glycerol. Fuel Process Technol 91(5):520–530

Kamwilaisak K, Wright PC (2012) Investigating laccase and titanium dioxide for lignin degradation. Energy Fuels 26(4):2400–2406

Karinen R, Vilonen K, Niemelä M (2011) Biorefining: heterogeneously catalyzed reactions of carbohydrates for the production of furfural and hydroxymethylfurfural. Chemsuschem 4(8):1002–1016

Khanna S, Goyal A, Moholkar VS (2012) Microbial conversion of glycerol: present status and future prospects. Crit Rev Biotechnol 32(3):235–262

Kim MG, Boyd G, Strickland R (1994) Adhesive properties of furfural-modified phenol-formaldehyde resins as oriented strandboard binders

Kim KH, Brown RC, Kieffer M, Bai X (2014) Hydrogen-donor-assisted solvent liquefaction of lignin to short-chain alkylphenols using a micro reactor/gas chromatography system. Energy Fuels 28(10):6429–6437

Kleinert M, Barth T (2008) Phenols from lignin. Chem Eng Technol 31(5):736–745

Kobayashi H, Ohta H, Fukuoka A (2012) Conversion of lignocellulose into renewable chemicals by heterogeneous catalysis. Catal Sci Technol 2(5):869–883

Kongjao S, Damronglerd S, Hunsom M (2010) Purification of crude glycerol derived from waste used-oil methyl ester plant. Korean J Chem Eng 27(3):944–949

Kongjao S, Damronglerd S, Hunsom M (2011) Electrochemical reforming of an acidic aqueous glycerol solution on Pt electrodes. J Appl Electrochem 41(2):215–222

Korpi H, Sippola V, Filpponen I, Sipilä J, Krause O, Leskelä M, Repo T (2006) Copper-2,2′-bipyridines: catalytic performance and structures in aqueous alkaline solutions. Appl Catal A 302(2):250–256

Kruger JS, Cleveland NS, Zhang S, Katahira R, Black BA, Chupka GM, Lammens T, Hamilton PG, Biddy MJ, Beckham GT (2016) Lignin depolymerization with nitrate-intercalated hydrotalcite catalysts. ACS Catalysis 6(2):1316–1328

Kumari N, Olesen JK, Pedersen CM, Bols M (2011) Synthesis of 5-bromomethylfurfural from cellulose as a potential intermediate for biofuel. Eur J Org Chem 7:1266–1270

Kunkes EL, Soares RR, Simonetti DA, Dumesic JA (2009) An integrated catalytic approach for the production of hydrogen by glycerol reforming coupled with water-gas shift. Appl Catal B 90(3):693–698

Lakshmi Ch V, Ravuru U, Kotra V, Bankupalli S, Prasad R (2009) Novel route for recovery of glycerol from aqueous solutions by reversible reactions

Lalitha K, Sadanandam G, Kumari VD, Subrahmanyam M, Sreedhar B, Hebalkar NY (2010) Highly stabilized and finely dispersed Cu_2O/TiO_2: a promising visible sensitive photocatalyst for continuous production of hydrogen from glycerol: water mixtures. J Phys Chem C 114 (50):22181–22189

Lam E, Chong JH, Majid E, Liu Y, Hrapovic S, Leung AC, Luong JH (2012) Carbocatalytic dehydration of xylose to furfural in water. Carbon 50(3):1033–1043

Lange JP, van der Heide E, van Buijtenen J, Price R (2012) Furfural—a promising platform for lignocellulosic biofuels. Chemsuschem 5(1):150–166

Leung DY, Wu X, Leung M (2010) A review on biodiesel production using catalyzed transesterification. Appl Energy 87(4):1083–1095

Li Q-S, Su M-G, Wang S (2007) Densities and excess molar volumes for binary glycerol + 1-propanol, + 2-propanol, + 1,2-propanediol, and + 1,3-propanediol mixtures at different temperatures. J Chem Eng Data 52(3):1141–1145

Lima S, Antunes MM, Pillinger M, Valente AA (2011) Ionic liquids as tools for the acid-catalyzed hydrolysis/dehydration of Saccharides to furanic aldehydes. Chem Cat Chem 3(11):1686–1706

Lin Y-C (2013) Catalytic valorization of glycerol to hydrogen and syngas. Int J Hydrogen Energy 38(6):2678–2700

Liu Q, Tian M, Ding T, Shi R, Feng Y, Zhang L, Chen D, Tian W (2007) Preparation and characterization of a thermoplastic poly (glycerol sebacate) elastomer by two-step method. J Appl Polym Sci 103(3):1412–1419

Liu G, Wu J, Zhang IY, Chen Z-N, Li Y-W, Xu X (2011) Theoretical studies on thermochemistry for conversion of 5-chloromethylfurfural into valuable chemicals. J Phys Chem A 115 (46):13628–13641

López JÁS, MdlÁM Santos, Pérez AFC, Martín AM (2009) Anaerobic digestion of glycerol derived from biodiesel manufacturing. Biores Technol 100(23):5609–5615

Magnusson L-E, Anisimov MP, Koropchak JA (2010) Evidence for sub-3 nanometer neutralized particle detection using glycerol as a condensing fluid. J Aerosol Sci 41(7):637–654

Mamman AS, Lee JM, Kim YC, Hwang IT, Park NJ, Hwang YK, Chang JS, Hwang JS (2008) Furfural: hemicellulose/xylosederived biochemical. Biofuels, Bioprod Biorefin 2(5):438–454

Manosak R, Limpattayanate S, Hunsom M (2011) Sequential-refining of crude glycerol derived from waste used-oil methyl ester plant via a combined process of chemical and adsorption. Fuel Process Technol 92(1):92–99

Marcotullio G, De Jong W (2010) Chloride ions enhance furfural formation from D-xylose in dilute aqueous acidic solutions. Green Chem 12(10):1739–1746

Marshall A, Haverkamp R (2008) Production of hydrogen by the electrochemical reforming of glycerol–water solutions in a PEM electrolysis cell. Int J Hydrogen Energy 33(17):4649–4654

Mascal M, Nikitin EB (2008) Direct, high-yield conversion of cellulose into biofuel. Angew Chem 120(41):8042–8044

May A, Salvadó J, Torras C, Montané D (2010) Catalytic gasification of glycerol in supercritical water. Chem Eng J 160(2):751–759

Mikkonen KS, Heikkinen S, Soovre A, Peura M, Serimaa R, Talja RA, Helén H, Hyvönen L, Tenkanen M (2009) Films from oat spelt arabinoxylan plasticized with glycerol and sorbitol. J Appl Polym Sci 114(1):457–466

Modig T, Granath K, Adler L, Lidén G (2007) Anaerobic glycerol production by *Saccharomyces cerevisiae* strains under hyperosmotic stress. Appl Microbiol Biotechnol 75(2):289

Montero JM, Gai P, Wilson K, Lee AF (2009) Structure-sensitive biodiesel synthesis over MgO nanocrystals. Green Chem 11(2):265–268

Moreau C, Durand R, Peyron D, Duhamet J, Rivalier P (1998) Selective preparation of furfural from xylose over microporous solid acid catalysts. Ind Crops Prod 7(2):95–99

Nascimento JE, Barcellos AM, Sachini M, Perin G, Lenardão EJ, Alves D, Jacob RG, Missau F (2011) Catalyst-free synthesis of octahydroacridines using glycerol as recyclable solvent. Tetrahedron Lett 52(20):2571–2574

Nimlos MR, Qian X, Davis M, Himmel ME, Johnson DK (2006) Energetics of xylose decomposition as determined using quantum mechanics modeling. J Phys Chem A 110 (42):11824–11838

O'Neill R, Ahmad MN, Vanoye L, Aiouache F (2009) Kinetics of aqueous phase dehydration of xylose into furfural catalyzed by ZSM-5 zeolite. Ind Eng Chem Res 48(9):4300–4306

Pan Y, Wang X, Yuan Q (2011) Thermal, kinetic, and mechanical properties of glycerol-plasticized wheat gluten. J Appl Polym Sci 121(2):797–804

Pandey MP, Kim CS (2011) Lignin depolymerization and conversion: a review of thermochemical methods. Chem Eng Technol 34(1):29–41

Parzuchowski PG, Grabowska M, Jaroch M, Kusznerczuk M (2009) Synthesis and characterization of hyperbranched polyesters from glycerol-based AB2 monomer. J Polym Sci Part A: Polym Chem 47(15):3860–3868

Peng L, Lin L, Zhang J, Zhuang J, Zhang B, Gong Y (2010) Catalytic conversion of cellulose to levulinic acid by metal chlorides. Molecules 15(8):5258–5272

Perego C, Bosetti A (2011) Biomass to fuels: the role of zeolite and mesoporous materials. Microporous Mesoporous Mater 144(1):28–39

Pompeo F, Santori G, Nichio NN (2010) Hydrogen and/or syngas from steam reforming of glycerol. Study of platinum catalysts. Int J Hydrogen Energy 35(17):8912–8920

Qi X, Watanabe M, Aida TM, Smith RL (2010) Fast transformation of glucose and di-/polysaccharides into 5-hydroxymethylfurfural by microwave heating in an ionic liquid/catalyst system. Chemsuschem 3(9):1071–1077

Qi X, Guo H, Li L (2011) Efficient conversion of fructose to 5-hydroxymethylfurfural catalyzed by sulfated zirconia in ionic liquids. Ind Eng Chem Res 50(13):7985–7989

Qi X, Watanabe M, Aida TM, Smith RL (2012) Synergistic conversion of glucose into 5-hydroxymethylfurfural in ionic liquid–water mixtures. Biores Technol 109:224–228

Ragauskas AJ, Beckham GT, Biddy MJ, Chandra R, Chen F, Davis MF, Davison BH, Dixon RA, Gilna P, Keller M (2014) Lignin valorization: improving lignin processing in the biorefinery. Science 344(6185):1246843

Rennard DC, Kruger JS, Schmidt LD (2009) Autothermal catalytic partial oxidation of glycerol to syngas and to non-equilibrium products. Chemsuschem 2(1):89–98

Roberts V, Fendt S, Lemonidou AA, Li X, Lercher JA (2010) Influence of alkali carbonates on benzyl phenyl ether cleavage pathways in superheated water. Appl Catal B 95(1):71–77

Roberts V, Stein V, Reiner T, Lemonidou A, Li X, Lercher JA (2011) Towards quantitative catalytic lignin depolymerization. Chem—Eur J 17(21):5939–5948

Román-Leshkov Y, Chheda JN, Dumesic JA (2006) Phase modifiers promote efficient production of hydroxymethylfurfural from fructose. Science 312(5782):1933–1937

Rosatella AA, Simeonov SP, Frade RF, Afonso CA (2011) 5-Hydroxymethylfurfural (HMF) as a building block platform: biological properties, synthesis and synthetic applications. Green Chem 13(4):754–793

Saleh J, Tremblay AY, Dubé MA (2010) Glycerol removal from biodiesel using membrane separation technology. Fuel 89(9):2260–2266

Santacesaria E, Vicente GM, Di Serio M, Tesser R (2012) Main technologies in biodiesel production: state of the art and future challenges. Catal Today 195(1):2–13

Seo H-B, Yeon J-H, Jeong MH, Kang DH, Lee H-Y, Jung K-H (2009) Aeration alleviates ethanol inhibition and glycerol production during fed-batch ethanol fermentation. Biotechnol Bioprocess Eng 14(5):599

Sergeev AG, Webb JD, Hartwig JF (2012) A heterogeneous nickel catalyst for the hydrogenolysis of aryl ethers without arene hydrogenation. J Am Chem Soc 134(50):20226–20229

Serrano-Ruiz JC, Dumesic JA (2011) Catalytic routes for the conversion of biomass into liquid hydrocarbon transportation fuels. Energy Environ Sci 4(1):83–99

Serrano-Ruiz JC, Luque R, Sepulveda-Escribano A (2011) Transformations of biomass-derived platform molecules: from high added-value chemicals to fuels via aqueous-phase processing. Chem Soc Rev 40(11):5266–5281

Sharma Y, Singh B, Upadhyay S (2008) Advancements in development and characterization of biodiesel: a review. Fuel 87(12):2355–2373

Shi S, Guo H, Yin G (2011a) Synthesis of maleic acid from renewable resources: catalytic oxidation of furfural in liquid media with dioxygen. Catal Commun 12(8):731–733

Shi X, Wu Y, Li P, Yi H, Yang M, Wang G (2011b) Catalytic conversion of xylose to furfural over the solid acid/ZrO_2-Al_2O_3/SBA-$_{15}$ catalysts. Carbohyd Res 346(4):480–487

Sitthisa S, Resasco DE (2011) Hydrodeoxygenation of furfural over supported metal catalysts: a comparative study of Cu. Pd and Ni Catal Lett 141(6):784–791

Son S, Toste FD (2010) Non-oxidative vanadium-catalyzed C–O Bond Cleavage: application to degradation of lignin model compounds. Angew Chem Int Ed 49(22):3791–3794

Ståhlberg T, Fu W, Woodley JM, Riisager A (2011) Synthesis of 5-(Hydroxymethyl) furfural in Ionic liquids: paving the way to renewable chemicals. Chemsuschem 4(4):451–458

Sturgeon MR, O'Brien MH, Ciesielski PN, Katahira R, Kruger JS, Chmely SC, Hamlin J, Lawrence K, Hunsinger GB, Foust TD (2014) Lignin depolymerisation by nickel supported layered-double hydroxide catalysts. Green Chem 16(2):824–835

Sun W, Liu D-Y, Zhu H-Y, Shi L, Sun Q (2010) A new efficient approach to 3-methylindole: vapor-phase synthesis from aniline and glycerol over Cu-based catalyst. Catal Commun 12 (2):147–150

Taarning E, Osmundsen CM, Yang X, Voss B, Andersen SI, Christensen CH (2011) Zeolite-catalyzed biomass conversion to fuels and chemicals. Energy Environ Sci 4(3):793–804

Takagaki A, Ohara M, Nishimura S, Ebitani K (2010) One-pot formation of furfural from xylose via isomerization and successive dehydration reactions over heterogeneous acid and base catalysts. Chem Lett 39(8):838–840

Tan KT, Lee KT, Mohamed AR (2010) A glycerol-free process to produce biodiesel by supercritical methyl acetate technology: an optimization study via response surface methodology. Biores Technol 101(3):965–969

Tao F, Song H, Chou L (2010) Efficient process for the conversion of xylose to furfural with acidic ionic liquid. Can J Chem 89(1):83–87

Thring R (1994) Alkaline degradation of ALCELL® lignin. Biomass Bioenerg 7(1–6):125–130

Tizvar R, McLean DD, Kates M, Dubé MA (2009) Optimal separation of glycerol and methyl oleate via liquid–liquid extraction. J Chem Eng Data 54(5):1541–1550

Toledano A, Serrano L, Labidi J (2014) Improving base catalyzed lignin depolymerization by avoiding lignin repolymerization. Fuel 116:617–624

Tong X, Ma Y, Li Y (2010) Biomass into chemicals: conversion of sugars to furan derivatives by catalytic processes. Appl Catal A 385(1):1–13

Towey J, Soper A, Dougan L (2011) The structure of glycerol in the liquid state: a neutron diffraction study. Phys Chem Chem Phys 13(20):9397–9406

Union E (2009) Directive 2009/28/EC of the European Parliament and of the Council of 23 April 2009 on the promotion of the use of energy from renewable sources and amending and subsequently repealing Directives 2001/77/EC and 2003/30/EC. Off J Eur Union 5:2009

Vaidya PD, Rodrigues AE (2009) Glycerol reforming for hydrogen production: a review. Chem Eng Technol 32(10):1463–1469

Valliyappan T, Bakhshi N, Dalai A (2008a) Pyrolysis of glycerol for the production of hydrogen or syn gas. Biores Technol 99(10):4476–4483

Valliyappan T, Ferdous D, Bakhshi N, Dalai A (2008b) Production of hydrogen and syngas via steam gasification of glycerol in a fixed-bed reactor. Top Catal 49(1–2):59–67

Vassilev SV, Vassileva CG, Vassilev VS (2015) Advantages and disadvantages of composition and properties of biomass in comparison with coal: an overview. Fuel 158:330–350

Vigier KDO, Benguerba A, Barrault J, Jérôme F (2012) Conversion of fructose and inulin to 5-hydroxymethylfurfural in sustainable betaine hydrochloride-based media. Green Chem 14 (2):285–289

Voitl T, Rudolf von Rohr P (2008) Oxidation of lignin using aqueous polyoxometalates in the presence of alcohols. Chemsuschem 1(8–9):763–769

Wang W (2010) Thermodynamic analysis of glycerol partial oxidation for hydrogen production. Fuel Process Technol 91(11):1401–1408

Wang Z, Zhuge J, Fang H, Prior BA (2001) Glycerol production by microbial fermentation: a review. Biotechnol Adv 19(3):201–223

Wang X, Li M, Wang M, Wang H, Li S, Wang S, Ma X (2009) Thermodynamic analysis of glycerol dry reforming for hydrogen and synthesis gas production. Fuel 88(11):2148–2153

Wang P, Yu H, Zhan S, Wang S (2011) Catalytic hydrolysis of lignocellulosic biomass into 5-hydroxymethylfurfural in ionic liquid. Biores Technol 102(5):4179–4183

West RM, Liu ZY, Peter M, Gärtner CA, Dumesic JA (2008) Carbon–carbon bond formation for biomass-derived furfurals and ketones by aldol condensation in a biphasic system. J Mol Catal A Chem 296(1):18–27

White CA, Kennedy JF (1985) In: Higuchi T, Chang H-M, Kirk TK (eds) Recent advances in lignin biodegradation research. Uni Publishers Co., Japan

Wolfson A, Dlugy C, Shotland Y, Tavor D (2009) Glycerol as solvent and hydrogen donor in transfer hydrogenation–dehydrogenation reactions. Tetrahedron Lett 50(43):5951–5953

Woodford JJ, Dacquin J-P, Wilson K, Lee AF (2012) Better by design: nanoengineered macroporous hydrotalcites for enhanced catalytic biodiesel production. Energy Environ Sci 5 (3):6145–6150

Xi Y, Davis RJ (2010) Glycerol-intercalated Mg-Al hydrotalcite as a potential solid base catalyst for transesterification. Clays Clay Miner 58(4):475–485

Xiang X, He L, Yang Y, Guo B, Tong D, Hu C (2011) A one-pot two-step approach for the catalytic conversion of glucose into 2, 5-diformylfuran. Catal Lett 141(5):735–741

Xu C, Arancon RAD, Labidi J, Luque R (2014) Lignin depolymerisation strategies: towards valuable chemicals and fuels. Chem Soc Rev 43(22):7485–7500

Yaakob Z, Mohammad M, Alherbawi M, Alam Z, Sopian K (2013) Overview of the production of biodiesel from waste cooking oil. Renew Sustain Energy Rev 18:184–193

Yang W, Sen A (2010) One-step catalytic transformation of carbohydrates and cellulosic biomass to 2,5-dimethyltetrahydrofuran for liquid fuels. Chemsuschem 3(5):597–603

Ye J, Sha Y, Zhang Y, Yuan Y, Wu H (2011) Glycerol extracting dealcoholization for the biodiesel separation process. Biores Technol 102(7):4759–4765

Yong G, Zhang Y, Ying JY (2008) Efficient catalytic system for the selective production of 5-Hydroxymethylfurfural from glucose and fructose. Angew Chem 120(48):9485–9488

Zakrzewska ME, Bogel-Łukasik E, Bogel-Łukasik R (2010) Ionic liquid-mediated formation of 5-hydroxymethylfurfural. A promising biomass-derived building block. Chem Rev 111 (2):397–417

Zhang Z, Zhao ZK (2010) Microwave-assisted conversion of lignocellulosic biomass into furans in ionic liquid. Biores Technol 101(3):1111–1114

Zhang B, Tang X, Li Y, Xu Y, Shen W (2007) Hydrogen production from steam reforming of ethanol and glycerol over ceria-supported metal catalysts. Int J Hydrogen Energy 32(13):2367–2373

Zhang J, Zhuang J, Lin L, Liu S, Zhang Z (2012) Conversion of D-xylose into furfural with mesoporous molecular sieve MCM-41 as catalyst and butanol as the extraction phase. Biomass Bioenerg 39:73–77

Zhao H, Holladay JE, Brown H, Zhang ZC (2007) Metal chlorides in ionic liquid solvents convert sugars to 5-hydroxymethylfurfural. Science 316(5831):1597–1600

Zhao Y, Xu Q, Pan T, Zuo Y, Fu Y, Guo Q-X (2013) Depolymerization of lignin by catalytic oxidation with aqueous polyoxometalates. Appl Catal A 467:504–508

Zhou C-HC, Beltramini JN, Fan Y-X, Lu GM (2008) Chemoselective catalytic conversion of glycerol as a biorenewable source to valuable commodity chemicals. Chem Soc Rev 37 (3):527–549

Zhou C-H, Xia X, Lin C-X, Tong D-S, Beltramini J (2011) Catalytic conversion of lignocellulosic biomass to fine chemicals and fuels. Chem Soc Rev 40(11):5588–5617

Chapter 8
Catalytic Pyrolysis of Biomass

8.1 Introduction

A process which involved thermochemical conversion for the in situ upgrading the quality of the bio-oil (pyrolysis oil) in the pyrolysis reactor with the used of heterogeneous catalyst as a heat carrier is called as catalytic pyrolysis of biomass. The upgrading of bio-oil was carried out to minimize the undesirable properties such as high viscosity, high oxygen and water content, instability, corrosivity and low heating value of the bio-oil. In the industry of petrochemical that was utilised to convert heavy oil fractions into chemicals and also lighter fuels, heterogeneous catalysis is used widely. In pyrolysis of biomass, this concept is also transferred where, by coming in contact with the suitable catalyst, the heavy oxygenated volatiles were deoxygenated and then converted to chemicals and also lighter fuels. The heavy oxygenated volatiles were obtained from the decomposition of biomass (Bridgwater 1994).

Ultimately, production of liquid with properties that has been improved which can be utilised directly as a liquid fuel or feedstock in modern refineries, same as crude oil is the main goal of the pyrolysis process (Bridgwater 1994). In literature, the upgrading of downstream of the thermal (conventional) bio-oil has been studied widely. However, many problem has been created and it was proved to be an issue which was complicated. In contrast, it was believed that some of these problems can be solved by utilise the catalytic bio-oil. For the process of biomass pyrolysis, an ideal catalyst must produce high yield and quality of bio-oil, with low amount of water and oxygen, exhibit both thermal stability and resistance to deactivation, and the quantity of undesirable products in the bio-oil is minimize. A product with better chemical and physical properties can be produced by increase the heating value of the bio-oil which can be done by reduce the oxygenated compounds in the bio-oil. Oxygen in the form of CO_2, H_2O and CO were removed from the vapours of pyrolysis process. When compared to CO, the oxygen removal is preferred to be in the form of CO_2. This is because, in the case of formation of CO, to remove one

© Springer International Publishing AG 2017
S. Bagheri, *Catalysis for Green Energy and Technology*,
Green Energy and Technology, DOI 10.1007/978-3-319-43104-8_8

oxygen atom one carbon atom is required, whereas in the case of CO_2 formation, to remove two oxygen atoms only one carbon atoms are required.

However, when compared to H_2O formation, both CO and CO_2 formation are more preferred for the purpose of oxygen removal because the preservation of the hydrogen is required for the hydrocarbon-forming reactions (Williams and Horne 1995a). The ratio of C-H can be enhanced effectively by the removal of oxygen as CO_2, and this will lead to reduction of coke deposition. Extreme attention has been received by some research groups on the use of heterogeneous catalysis to improve the quality of the bio-oil. In order to develop this technology, few challenges which involve the lignocellulosic materials processing and also the optimization of advanced porous materials as effective bifunctional and monofunctional catalysts for the conversion of biomass to transportation fuels.

8.2 Catalysts for Biomass Catalytic Pyrolysis

Some catalytic materials have been investigated as candidate catalysts to be use in pyrolysis of biomass, such as base catalysts, zeolites, mesoporous/microporous hybrid materials which are doped with transition and noble metals, and also mesoporous materials with uniform size distribution of pore (SBA-15, MSU and MCM-41). These catalysts must be able to selectively favour the decarboxylation reactions, which will produce bio-oil that is high-quality and have low amounts of water and oxygen. The formation of oxygenated compounds which were undesirable such as acids, carbonyl and ketones must be prevented because they are known to be not acceptable for further co-processing and direct use of the bio-oil. Improvement in the hydrothermal stability of the catalyst are also required; therefore, investigation on behaviour of catalyst and resistance to deactivation of catalyst must be carried out for new catalysts optimization. In general, the process on development of catalyst includes controlling the formation of suitable catalyst particles and tailoring the basicity, acidity, porosity and interactions between support-metal of the materials of candidate catalysts.

8.2.1 Basic Catalysts

For the conversion of biomass to oils by pyrolysis process, solid acid catalysts are one of the predominately tested materials. However, in the catalytic pyrolysis, basic oxides such as ZnO and MgO also has been applied and demonstrated by Nokkosmaki et al. (2000) and Fabbri et al. (2007) respectively. To investigate the influence of the catalysts on the stability and also composition of the bio-oil, pyrolysis processes were carried out. Pyrolysis of pine sawdust using three different ZnO catalysts at the temperature of 600 °C, with residence time of 30 ms passing the pyrolysis of pine biomass vapours through a fixed bed of catalysts was demonstrated. They found that, for production of bio-oils, ZnO can be used as a mild catalyst, where the gas was increased only by 2 wt%, with a small reduction in the yield of liquid, but he catalytic oil stability was increased significantly.

When the temperature of the catalyst was increased, the reduction of formic acid and anhydrosugars formation can be seen. Both formic acid and anhydrosugars were not detected when the process was carried out at the usual temperature of the biomass pyrolysis which is at 500 °C (Nokkosmäki et al. 2000). In a Py-GC-MS system, the biomass pyrolysis vapours upgrading have been studied by Lu et al. and different catalytic capabilities can be observed by using different metal oxides as the materials for catalysts. The acids can be eliminated and the levels of anhydrosugars and phenols can be reduced significantly by using CaO. The formation of several light compounds, hydrocarbons and also cyclopentanones can also be increased with CaO. Various hydrocarbons formation was observed with Fe_2O_3 and ZnO was found to be a mild catalyst. Using thermogravimetric analysis-fourier transform infrared spectroscopy analysis, the CaO has been studied by Wang et al. and it was found that the hydrocarbons formation were promoted and it can effectively reduce the acids (Wang et al. 2010). Finally, in a system of pyrolysis-gas chromatography-microwave induced plasma-atomic emission detector (Py-GC-MIP-AED), the catalytic pyrolysis of pine sawdust using various metal oxides has been investigated by Torri et al. and after catalysis the reduced of heavy compounds can be observed at the expense of bio-oil yields. From their study, it was found that the most interesting materials were CuO, ZnO and mixed metal oxide catalysts, where the CuO will obtained the highest yields in semi-volatile compounds and ZnO will results in reduction of heavy fraction proportion with a limited decrease in the yield of bio-oil (Torri et al. 2010).

For the upgrading process of biomass pyrolysis vapours, TiO_2 Anatase, TiO_2 Rutile and ZrO_2/TiO_2 catalysts and their modified counterparts that has been incorporated with Pd, Ru and Ce has been utilised by Lu et al. in a Py-GC-MS system. Promising effects to favoured the reduction of sugars and aldehydes, while increasing the acids, cyclopentanones and ketones was exhibited by TiO_2 Rutile catalysts especially the counterpart with Pd-containing. Besides that, it was also promising to convert the oligomers of lignin-derived to monomeric phenols. While, the yields of acid and phenol can be reduced remarkably, increased the light linear ketones, cyclopentanones and hydrocarbons, and eliminated sugars can be observed with the use of ZrO_2/TiO_2 catalysts (Lu et al. 2010). The worked with basic materials such as MgO, TiO_2 and ZrO_2/TiO_2 has been carried out by a research group of Chemical Process Engineering Research Institute (CPERI).

When compared to acidic zeolites, different pathways of deoxygenation in pyrolysis of biomass has been shown by these catalysts tested. Specifically, high yields of cyclopentanones, cyclopentenones and ketones within liquid products and high yields of CO_2 were achieved. The mechanisms on the production of CO_2 have been presented elsewhere (Stefanidis et al. 2011). In short, aldol ketonization and condensation reactions are prevalent, mainly leading to the conversion of aldehydes to ketones and condensation of acids and formation of CO_2 as the main by-product (Deng et al. 2008; Gaertner et al. 2009). Moreover, these condensation reactions involved carbon-carbon coupling reactions that will condensate acid molecules and smaller aldehyde into larger molecules that have almost similar molecular weight to those liquid transportation fuels. In this case, so far, an interesting alternative to the acidic route using basic catalysis has been studied and achieved with MgO.

Catalytic pyrolysis of pinewood sawdust using different basicity of inorganic additives have been performed by Chen et al. by microwave heating. The composition and yields of the pyrolysis products have been altered by all of the eight different additives. However, significant decreased in the gases yield and increased in the solid product yield have been caused by them. There is no dramatic change in the liquid products yield. Gaseous products were evolve earlier using these different eight additives. From pyrolysis, the gases produced mainly of CO, CO_2, CH_4 and H_2. While, the formation of H2 in the pyrolytic gases is favoured by using alkaline sodium compounds such as Na_2CO_3, Na_2SiO_3 and NaOH (Chen et al. 2008). The pyrolysis of chlorella algae catalysed by Na_2CO_3 has been demonstrated by Babich et al. (2011). When compared the non-catalytic pyrolysis at the same temperature, decrease in bio-oil yield and increase in gas yield was shown in the presence of Na_2CO_3, but lower acidity and higher heating value can be observed for the pyrolysis oil produced from the catalytic pyrolysis (Babich et al. 2011). When ammonia was reacted with zeolites at elevated temperatures, zeolites which are strongly basic were synthesized, providing unique selectivity and activity for base-catalyzed reactions (Ernst et al. 2000). For production of hydrocarbons from biomass through catalytic pyrolysis, this basic zeolites such as amine-substituted ZSM-5 has been suggested as one of the promising candidate catalysts (Carlson et al. 2008, 2009). The deoxygenation of benzaldehyde on basic NaX and CsNAX zeolites have been investigated by Peralta et al. (2009). Using highly basic catalyst with excess of Cs, direct decarbonylation of benzene from benzaldehyde can be readily promoted. In parallel, formation and decomposition of toluene also can occurred through the condensation of surface products (Peralta et al. 2009).

8.2.2 Microporous Acidic Catalysts

In oil refineries, microporous acidic catalysts that are known to catalyse the carbon-carbon bonds scission of heavier oil fractions has been used widely. For the conversion of heavier to lighter oxygenates, similar mechanism is required for biomass pyrolysis. With respect to this, the used of acidic zeolite catalysts for catalytic biomass pyrolysis have been studied by several groups, while for the upgrading of the biomass pyrolysis vapours the main zeolite that has been studied is the ZSM-5 (Aho et al. 2007; Williams and Horne 1995b; Thring et al. 2000; Atutxa et al. 2005; Horne and Williams 1996b; Lappas et al. 2002). The effects of the ZSM-5 zeolite on the biomass pyrolysis have been investigated by Williams and Horne in a series of studies. After catalysis, it was found that a premium grade gasoline-type fuel was produced by significantly reduced the oxygenated species in the bio-oil and also increased their aromatic species. After been analysed in details, high concentrations of economically valuable chemicals can be found in the upgraded oil. Besides that, when the temperature of the catalyst bed was increased, concentration of hazardous, biologically active species of polycyclic aromatic were also increased (Horne and Williams 1994).

At lower temperatures of the catalyst bed, the removal of oxygen from the vapours of pyrolysis is mainly as H_2O, while at higher temperature they mainly removed as CO_2 and CO. with increase in the temperature of the catalyst bed, the shift towards species with lower molecular weight also can be observed (Williams and Horne 1994). The aromatic hydrocarbons derived from the catalytic biomass pyrolysis of bio-oil using HZSM-5 have been analysed in details and it shows that the single aromatic ring species produced are mainly toluene, alkylated benzenes and also benzenes. The polycyclic aromatic hydrocarbons (PAHs) were mainly fluorene, phenanthrene, naphthalene and also their alkylated homologues (Williams and Horne 1995a). During zeolite catalysis, there is only limited understanding on the reactions due to the complex nature of the components in the bio-oil. To simplify the problem, the reaction of oxygenated biomass pyrolysis model compounds by using a ZSM-5 catalyst have been investigated by Williams and Horne. At the temperature of 300–350 °C, catalytically methanol can be converted to hydrocarbon products, meanwhile higher temperatures are required for anisole, cyclopentanone and furfural.

In the upgrading of the oxygenated compounds, the formation of coke can be reduced when the temperature of the catalysts bed was increasing, with the exception of anisole. The optimum temperature for catalysis was at 500–550 °C. However, at that temperature, yields of coke gave by anisole were high, suggesting that the major coking components present in vapours of biomass-derived pyrolysis are the phenolic compounds. For anisole, their oxygen bond was appeared to be refractory to the catalytic upgrading, leading to formation of phenolic compounds in large quantities and the yields of aromatic hydrocarbons is low (Horne and Williams 1996a). The transformation of oxygenated bio-oil model compounds have been studied by Gayubo et al. (2004a, b) using HZSM-5 zeolite. They found that, at lower temperatures via dehydration from olefins alcohols was transformed to aromatic hydrocarbon and paraffins above 35 °C, while at 250 °C it was transformed to higher olefins.

Low reactivity was exhibited by phenol with the use of HZSM-5 zeolite. Other than generating thermal coke, low reactivity was also exhibited by 2-methoxyphenol. Good agreement can be observed between this observations on phenolic compounds and the observations of Williams and Horne (Horne and Williams 1996a). For both phenols and alcohols, low deactivation rate by deposition of coke can be observed. Formation of coke deposition that will contribute to capacity for oligomerization can be observed with acetaldehyde. Besides that, acetaldehyde also had a low reactivity to hydrocarbons. Through decarboxylation, transformation of acetic acid and ketones occurred and to a lesser extent, dehydration can occur. Similar to the reaction scheme of alcohols, aromatics and olefins were transformed at the temperature of above 400 °C. Significant amount of coke were generated more than in the corresponding process for alcohols and the formation of olefins which were the intermediate products of the reaction scheme were increased (Gayubo et al. 2004a, b).

In a later study, the transformation of bio-oil model compounds mixtures has been investigated by Gayubo et al. on the HZSM-5 zeolite and it was found that the results agree well with the results that used pure compounds. However, when compared to pure compounds, mixtures of model compound have synergistic effects due to the high reactivity of some of the products of primary pyrolysis. The deposition of coke results in severe deactivation of catalyst, which was enhanced by the thermal coke from the acetaldehyde degradation. The tendency of 2-methoxyphenol degradation to coke can be enhance in the presence of furfural. Thus, it can be concluded that to reduce the formation of coke, the feasibility of upgrading process of the bio-oil into hydrocarbons needs the previous separation of certain bio-oil's components for example furfural, oxyphenols, aldehydes and their derivatives (Gayubo et al. 2005). The bio-oil upgrading using HZSM-5 have been studied by Vitolo et al. (2001) and the catalytic activity was tentatively attributed to its acidic sites through the mechanism of carbonium ion, promote decarboxylation, decarbonylation and deoxygenation of the constituents of the oil as well as oligomerization, isomerization, alkylation, aromatization and also cyclization. The behaviour of the zeolite also has been investigated in repeated cycles of upgrading-regenerating. After the fifth cycle of upgrading-regenerating, irreversible poisoning of material was observed. The disappearance of a significant amount of acid sites especially the stronger ones might be the reason of the catalyst deactivation (Vitolo et al. 2001).

It is known that the catalytic upgrading of the pyrolysis vapours can reduce the yield of bio-oil and increase the yields of coke, water and also gas. When the catalyst came in contact with the feed of biomass, the biomass residue can be reduced and it have been reported by some groups. At the temperature of 400 °C, the effects of HZSM-5 zeolite on the in situ sawdust pyrolysis have been studied by Atutxa et al. (Gayubo et al. 2004a; Atutxa et al. 2005). When the mass of the catalyst was increased, the yields of the gas also increase and the yield of liquid will decrease. Slight reduced in the char formation and increased in the proportion of CO over CO_2 also can be observed. The reduction in the total product of liquid through the transformation of gases and light liquid product from heavy liquid fraction was reported by Atutxa et al. (2005). When compared to the heavy fraction, the light fraction was more severely deoxygenated, which is evidence when the lighter fraction compounds have higher global reactivity, especially of acetic acid and alcohols. Good agreement between this work and the work on bio-oil model compounds done by Williams and Horne (Horne and Williams 1996a) and Gayubo et al. (2004a) can be observed.

Less viscous, less oxygenated, more stable and less corrosive oil can be produced from catalytic pyrolysis (Atutxa et al. 2005). Three different kinds of biomass were pyrolyzed using the HZSM-5 zeolite in a study by Wang et al. and it have been found that the maximum weight loss rate was increased when the HZSM-5 is presence in the biomass (Chen et al. 2008). Using fluidized bed, the pyrolysis of corncob with the HZSM-5 catalyst have been studied by Zhang et al. From this study, it can be observed that the catalyst can cause an increase in the coke, H_2O and non-condensable gas yields and also decrease in the fraction of heavy oil (Zhang et al. 2009). The number of aromatic hydrocarbons in the fraction of oil was increased and all the other types of compounds was decreased when the HZSM-5 was used. When compared to non-catalytic oil (40.28%), the catalytic oil have lower content of oxygen (14.69%) (Zhang et al. 2009). With the use of ZSM-5, selectivity towards aromatics and its ability on reduction of oxygenates also were demonstrated by the other groups. Effects on using different catalytic materials for conversion of representative compounds of bio-oil model have been studied by Samolada et al. (2000). And it has been concluded that the undesirable carbonyls have been completely converted to hydrocarbon by HZSM-5 zeolite, simultaneously, fraction of organic liquid was loss and the H_2O was increased dramatically.

In the system of pyrolysis-gas chromatography-mass spectrometry (Py-GC-MS), upgrading of the vapours from cassava rhizome pyrolysis have been investigated by Pattiya et al. using various catalysts (Pattiya et al. 2008). It has been concluded that, ZSM-5 was the most active catalyst and the formation of phenols, acetic acid and aromatic hydrocarbons can be increased using this catalyst. The carbonyls and other oxygenated lignin-derived compounds can also be reduced (Pattiya et al. 2008). Lastly, the effects of mixtures of silica alumina and HZSM-5 towards the distribution of product during the conversion of bio-oil from maple wood have been studied by Adjaye et al. (1996). It can be observed that the organic liquid produced with HZSM-5 consisted mainly of aromatic hydrocarbons, whereas with silica-alumina, aliphatic hydrocarbons was produced. Formation of coke can be reduced and the yields of gas and organic liquid product can be increased when the HZSM-5 was added to silica-alumina. The products of hydrocarbon can be changed to aromatic from aliphatic with gradual increase of HZSM-5 in the mixture. From these observation, it can be suggested that when compared to less acidic silica-alumina, the hydrogen transfer is more effective with the used of HZSM-5 catalyst.

During upgrading, the aliphatic formed will be converted into the thermodynamically favoured aromatics. When compared to aliphatic hydrocarbons, the H–C ratio of aromatics is lower. Due to its low content of hydrogen, overall hydrocarbon formation from pyrolysis oil will be reduced, therefore more hydrogen has to be put into the aliphatics (Adjaye et al. 1996). The catalytic upgrading of lignin pyrolysis vapours by using ZSM-5 zeolite as catalyst have been studied by several groups. Lignin is one of the by-product of paper pulp mills and is also the main components of lignocellulosic biomass. In a fixed bed reactor, the catalytic pyrolysis of lignin which was solubilized in acetone with HZSM-5 have been studied by Thring et al. and it can be observed that the yields of gasoline range hydrocarbons for example toluene, xylene and benzene were high. In a system of Py-GC-MS, increased in the aromatic hydrocarbons production during the lignin pyrolysis was also observed by Boateng and Mullen in the presence of HZSM-5 catalyst.

The increased in the production of aromatic hydrocarbon can be due to the effectiveness of the catalyst which enhanced the depolymerization that converted and released the aliphatic linkers of the lignin into olefins, and the by aromatization it was converted to aromatic compounds. They also observed that, the release of simple phenols from lignin decomposition that will degraded to coke will results in deposition of coke, and then deactivation of catalyst can occurs (Mullen and Boateng 2010). Lignin pyrolysis by using five different catalyst have been studied by Jackson et al. and it can be concluded that the best catalyst to be used for deoxygenated liquid fraction production was the HZSM-5 catalyst (Jackson et al. 2009). Metals have been incorporated in the framework of zeolite by some research groups in an effort to improve the catalytic effect of ZSM-5 on the pyrolysis vapours upgrading. According to Czernik and French (2010) the mode of oxygen rejection was affected by the presence of transition metals, by producing of less H_2O and more CO_2, making in that way more hydrogen were available to be incorporated into the hydrocarbons. Using a system of Py-GC-MS, pyrolysis of xylan have been performed by Zhu et al. (2010) with addition of metal-impregnated iron(Fe)/HZSM-5 and Zn/HZSM-5 catalysts and HZSM-5 catalyst. From the results, it was found that for the production of higher hydrocarbons content, the metal-impregnated HZSM-5 catalysts were more effective due to their ability in reduction of oxygenated compounds. Performance in hydrocarbon production of laboratory synthesized and also commercial catalysts have been evaluated by Czernik and French (2010) and it can be concluded that ZSM-5 group was the best performing catalysts, while the highest yields of hydrocarbons can be achieved with Fe, Ni, and Ga substituted ZSM-5.

Less activity of deoxygenation was exhibited by zeolites with larger pores. The catalytic effects of Ga/HZSM-5, HZSM-5 and H–Y zeolite have been compared by Park et al. (2007) and it can be concluded that better selectivity to aromatic hydrocarbons and more bio-oil can be produced with Ga/HZSM-5 when compared to HZSM-5. While, HZSM-5 was more efficient in upgrading of bio-oil when compared to H–Y zeolite. Other acidic materials and microporous zeolites also have been studied in the literature. During the bio-oil conversion using silica-alumina, silicalite, H–Y zeolite, HZSM-5 zeolite and H-mordenite zeolite, Adjaye and Bakhshi (1995) observed that the less acidic silica-alumina and non-acidic silicalite were less effective for conversion of hydrocarbons from bio-oil when compared to acidic zeolites, and the highest hydrocarbons yields can be obtained with the use of HZSM-5 zeolite catalyst. Production of more aromatic than aliphatic hydrocarbons can be produced with H-mordenite and HZSM-5 zeolites catalysts, while more aliphatic than aromatic hydrocarbons can be produced with silicalite, silica-alumina and H-Y catalysts (Zhang et al. 2009; Adjaye and Bakhshi 1995).

In a fluidized bed reactor, the pyrolysis of pinewood with the use of proton forms of beta, mordenite and Y-ZSM-5 zeolites have been carried out by Aho et al. (2007) and it was found that the bio-oil chemical composition were depends on the structure of the utilised zeolite. Lower selectivity to alcohols and acids, higher selectivity to ketones and highest yields of liquid products can be gave by ZSM-5 zeolite, while the PAHs formation can be effectively minimized by mordenite zeolite (Aho et al. 2008). For the purpose of catalytic pyrolysis of corn stalks, different structures of zeolite such as H–Y, USY and ZSM-5 have been studied by Uzun and Sarioglu (2009). They observed the lowest oil yield with the USY and the highest with the ZSM-5. With USY catalyst, highest amounts of aromatics also can be observed. The aliphatic hydrocarbons can be increase with the H–Y zeolite (Uzun and Sarioğlu 2009). In a system of Py-GC-MS, the conversion of aromatics from biomass compounds using beta, Y and ZSM-5 zeolites have been studied by Carlson et al. (2008, 2009) and found that the production of aromatic compounds was favour over formation of coke when the ratios of catalyst to feed was high by avoiding the thermal decomposition reactions which are undesirable in the homogeneous phase. Volatile oxygenates such as acetic acid, furan-type compounds and hydroxyacetaldehyde were formed when the ratios of catalyst to feed was lower. The least formation of coke and highest yields of aromatic was obtained with ZSM-5 zeolite (Carlson et al. 2008, 2009).

For the upgrading of biomass fast pyrolysis vapours, HY and HZSM-5 zeolites together with the other three mesoporous materials have been studied by Lu et al. (Lu et al. 2009). For the pyrolysis vapours deoxygenation, the two microporous zeolites have been proved to be effective and abundant of aromatic hydrocarbons were formed after the catalysis together with some PAHs (Lu et al. 2009). In a fixed bed reactor, catalytic pyrolysis of bamboo was performed by Qi et al. with NaY zeolite catalyst and it can be observed after the catalysis the yields of the bio-oil were increased and consisted mainly of carbonylic and carboxylic compounds (Qi et al. 2006). The main component of the catalytic bio-oil was acetic acid and when compared to thermal pyrolysis oil, its content was higher with catalytic pyrolysis, whereas the carbonylic compounds content was makedly lower (Qi et al. 2006). In the upgrading of pinewood pyrolysis vapours, the influence of Y, beta and ferrierite zeolites and their counterparts of Fe-modified have been studied by Aho et al. They concluded that the most active catalyst in the deoxygenation reactions was beta zeolite, then followed by Y and lastly ferrierite. In the presence of Fe-modified zeolites, the amounts of methoxy-substituted phenols in the bio-oil decreased, while the methyl-substituted increased (Aho et al. 2010). The influence of acidity of the beta zeolite on the biomass catalytic pyrolysis also have been studied by Aho et al. They observed that the less acidic zeolites tend to form more organic oil and less polyaromatic hydrocarbons and H_2O than zeolites with stronger acidity (Aho et al. 2007).

8.2.3   Mesoporous Acidic Catalysts

There are some serious drawbacks were observed with zeolitic materials, for example undesirable PAHs formation, rapid deactivation of catalyst, and significant increase in the H_2O production. Recently, much attention has been attracted to catalysts that have pore size which was larger than that of the zeolites in order to overcome these problems. This is because, they are expected to allow the products of pyrolysis with larger molecules to enter, then reformulate and at the end exit the matrix of the catalyst with less chance of pores blocking and deposition of coke (Pattiya et al. 2008). As reported in the literature, among the mesoporous acidic materials catalyst, the most focus has been received by the MCM-41 catalyst. MCM-41 was discovered in 1992 and it is the main representative of the mesoporous molecular sieve of M41S family. High surface area (>1000 m^2/g) was exhibited by MCM-41 and it possesses a uniform mesopores hexagonal array which dimensions can be varied from 1.4 to <10 nm in size.

Studies on SBA-15 catalysts have been carried out as well. When compared to MCM-41-type materials, SBA-15 was more hydrothermally and thermally stable. It can be due to their thicker walls, large monodispersed mesopores and long-range order (Adam et al. 2006). For the purpose of biomass catalytic pyrolysis, the mesoporous Al-MCM-41 material have been evaluated by Samolada et al. and it was found that its hydrothermal stability was poor, therefore to be used in the process of pyrolysis the further optimization of this material is needed (Samolada et al. 2000). The acidity effects and stability of steam of Al-MCM-41 have been studied by Iliopoulou et al. (2007). They found that the conversion of pyrolysis vapours to coke and gas was favoured with higher number of acid sites, whereas beneficial effect for the liquid organic product production can be obtained with lower number of acid sites. The number of acid sites and their surface area can be reduced by 40–60% when moderate steaming was applied to the materials. However, for the biomass pyrolysis vapours upgrading, the steamed samples were still active (Iliopoulou et al. 2007).

In a system of Py-GC-MS, the spruce wood biomass pyrolysis has been studied by Adam et al. with four different Al-MCM-41-type catalysts and modified Al-MCM-41 catalysts. After catalysis, they found that the high molecular mass phenols yields were decreased and the furans and acetic acid yields were increased. Overall, there was increased in the yields of the phenols, and there was also a slight increase in the yields of hydrocarbon. The yields of H$_2$O and acetic acid were reduced with the enlargement of the pores of the catalyst, and in general production of higher molecular mass products were led by the introduction of the Cu cation and also the enlargement of the pore (Adam et al. 2005). In another study, in a fixed bed reactor for upgrading of biomass pyrolysis vapours, the effects of an aluminium SBA-15, four Al-MCM-41 catalysts with Si–Al ratio of 20, and a pure siliceous SBA-15 have been investigated by Adam et al. The Al-MCM-41 materials that have been modified by introduction of Cu cation into their structure and also enlargement of their pore also was studied. They found that after catalysis with all materials, the yields of H$_2$O and gas were increased. The yields of acid and carbonyl were decreased, whereas the yields of PAHs, phenol and also hydrocarbon were increased. It can also be observed that the presence of compounds which are undesirable in the bio-oil were reduced by all the catalysts. The quality of the bio-oil was seemed to be effected by the enlargement of the pore of Al-MCM-41. On the other hand, with the introduction of the Cu cation, beneficial effect and increased in the production of desirable products can be observed.

A very high content of desirable products in the bio-oil produced can be obtained when aluminium was incorporated into the framework of the SBA-15. In a fixed bed reactor for pyrolysis of biomass, three materials of Al-MCM-41 with different ratios of Si–Al and three metal containing (Fe, Zn and Cu)-Al-MCM-41 materials have been evaluated by Antonakou et al. (2006). When compared to the non-catalytic pyrolysis, the bio-oil production and its organic fraction were decreased, whereas the coke formation was increased. The production of phenolic compounds was increased with all the catalysts. For production of phenols, Fe-Al-MCM-41 was the best catalyst, followed by Cu-Al-MCM-41 and the lowest was Si-Al-MCM-41. A decreased in the fractions of carbonyl, acids and heavy compounds which are undesirable have been observed with almost all of the tested catalyst materials. They found that the composition and yields of the product was positively affected by lower ratios of Si–Al (Antonakou et al. 2006).

For catalytic pyrolysis of biomass, the parent materials of Al-MCM-41 and their metal modified with Si–Al ratio of 20 have also been studied by Nilsen et al. (2007). Better quality of bio-oil with respect to yields of phenol can be observed with all materials. The phenol yields were lower when the Zn-Al-MCM-41 catalysts was used, however it gave best results with respect to production of coke. For catalytic pyrolysis of biomass, the MSU materials also have been tested. In a fixed bed reactor for upgrading of biomass pyrolysis vapours, two aluminosilicate MSU-S materials with different structure of mesopore have been assembled by Triantafyllidis et al. and then they were compared with the conventional materials of Al-MCM-41 (Triantafyllidis et al. 2007). When compared to non-catalytic pyrolysis and the conventional Al-MCM-41 materials, the organic phase of the bio-oil were significantly reduced and led to higher yields of coke with the used of MSU-S materials. The selectivity of MSU-S materials towards heavy fractions and PAHs were also improved, but negligible amounts of alcohols, acids, carbonyls and very few phenols were produced (Triantafyllidis et al. 2007).

In a system of Py-GC-MS, the materials of Al-MCM-41 and Al-MSU-F have been studied by Pattiya et al. and it was found that potential improvement of the bio-oil was demonstrated in therms of heating value and initial viscosity. When compared to ZSM-5 zeolite, they tend to reduce the lignin-derived compounds which are deoxygenated and produce hydrocarbons. On the downside, it can be observed that the yields of acetic were increased (Pattiya et al. 2008). For the upgrading of lignin pyrolysis vapours, the comparison between HZSM-5 zeolite and mesoporous Al-MCM-41 have been done by Jackson et al. and it can be observed that the Al-MCM-41 behaved similarly to the HZSM-5, even though it was not as effective in deoxygenating the liquid phase and gave more naphthalenes than simple aromatics (Jackson et al. 2009). Using a system of Py-GC-MS, the materials of Al-SBA-15 and siliceous SBA-15 have been studied by Lu et al. and when compared to the acidic zeolites, they were found to be mild catalysts. They favoured the formation of furfural, furan, light phenols, acetic acid and the other light compounds of furan and decreased the yields of many light carbonyls, heavy furans and also heavy phenols. When compared to siliceous SBA-15, the materials of Al-SBA-15 were more active and their catalytic effects can be enhanced by decreasing the ratios of the Si–Al of the catalyst (Lu et al. 2009).

Lately, the synthesized of the mesoporous zeolites have been carried out to be studied for the catalytic pyrolysis of biomass. By using commercial beta and ZSM-5 zeolites, ordered mesoporous aluminosilicates (MMZ) have been synthesized by Lee et al. and then applied them to the pyrolysis of woody biomass. It was found that the synthesized material has an excellent hydrothermal stability and also exhibits a well-developed mesoporosity. It was also found to be promising catalysts to be used for bio-oil upgrading, due to both its yields of organic fraction and also selectivity to phenolics, and also reduction of compounds which are undesirable such as oxygenates. Moreover, through the cycles of regenerating-upgrading, the high catalytic activities were maintained by the synthesized catalysts, whereas the catalytic activity of Al-MCM-41 was significantly decreased (Lee et al. 2008).

A mesoporous mordenite framework inverted (MFI) zeolite have been synthesized by Park et al. and its catalytic activity was compared with mesoporous material from HZSM-5 (MM-HZSM-5) and conventional HZSM-5. Best activity in terms of aromatization and deoxygenation was exhibited by the mesoporous MFI zeolite. In particular, high selectivity towards valuable BTX hydrocarbons were observed, but the yields of the organic fraction was decreased. When gallium was incorporated into the mesoporous MFI zeolite, the organic fraction of bio-oil and resistance to deposition of coke were increased and less cracking of the pyrolytic vapours were observed. The selectivity of aromatics depends on the amounts of gallium that were incorporated into the mesoporous MFI zeolite. The consequence enhancement of the selectivity for BTX aromatics and the bifunctional mechanism can be optimized with addition of appropriate amount of gallium, whereas negative effect on aromatics formation can be brought with excess gallium because of more protons loss (Park et al. 2010).

References

Adam J, Blazso M, Meszaros E, Stöcker M, Nilsen MH, Bouzga A, Hustad JE, Grønli M, Øye G (2005) Pyrolysis of biomass in the presence of Al-MCM-41 type catalysts. Fuel 84(12):1494–1502

Adam J, Antonakou E, Lappas A, Stöcker M, Nilsen MH, Bouzga A, Hustad JE, Øye G (2006) In situ catalytic upgrading of biomass derived fast pyrolysis vapours in a fixed bed reactor using mesoporous materials. Microporous Mesoporous Mater 96(1):93–101

Adjaye JD, Bakhshi N (1995) Production of hydrocarbons by catalytic upgrading of a fast pyrolysis bio-oil. Part I: conversion over various catalysts. Fuel Process Technol 45(3):161–183

Adjaye JD, Katikaneni SP, Bakhshi NN (1996) Catalytic conversion of a biofuel to hydrocarbons: effect of mixtures of HZSM-5 and silica-alumina catalysts on product distribution. Fuel Process Technol 48(2):115–143

Aho A, Kumar N, Eränen K, Salmi T, Hupa M, Murzin DY (2007) Catalytic pyrolysis of biomass in a fluidized bed reactor: influence of the acidity of H-beta zeolite. Process Saf Environ Prot 85(5):473–480

Aho A, Kumar N, Eränen K, Salmi T, Hupa M, Murzin DY (2008) Catalytic pyrolysis of woody biomass in a fluidized bed reactor: influence of the zeolite structure. Fuel 87(12):2493–2501

Aho A, Kumar N, Lashkul A, Eränen K, Ziolek M, Decyk P, Salmi T, Holmbom B, Hupa M, Murzin DY (2010) Catalytic upgrading of woody biomass derived pyrolysis vapours over iron modified zeolites in a dual-fluidized bed reactor. Fuel 89(8):1992–2000

Antonakou E, Lappas A, Nilsen MH, Bouzga A, Stöcker M (2006) Evaluation of various types of Al-MCM-41 materials as catalysts in biomass pyrolysis for the production of bio-fuels and chemicals. Fuel 85(14):2202–2212

Atutxa A, Aguado R, Gayubo AG, Olazar M, Bilbao J (2005) Kinetic description of the catalytic pyrolysis of biomass in a conical spouted bed reactor. Energy Fuels 19(3):765–774

Babich I, Van der Hulst M, Lefferts L, Moulijn J, O'Connor P, Seshan K (2011) Catalytic pyrolysis of microalgae to high-quality liquid bio-fuels. Biomass Bioenerg 35(7):3199–3207

Bridgwater A (1994) Catalysis in thermal biomass conversion. Appl Catal A 116(1–2):5–47

Carlson TR, Vispute TP, Huber GW (2008) Green gasoline by catalytic fast pyrolysis of solid biomass derived compounds. Chemsuschem 1(5):397–400

Carlson TR, Tompsett GA, Conner WC, Huber GW (2009) Aromatic production from catalytic fast pyrolysis of biomass-derived feedstocks. Top Catal 52(3):241

Chen M-q, Wang J, Zhang M-x, Chen M-g, Zhu X-f, Min F-f, Tan Z-c (2008) Catalytic effects of eight inorganic additives on pyrolysis of pine wood sawdust by microwave heating. J Anal Appl Pyrol 82(1):145–150

Deng L, Fu Y, Guo Q-X (2008) Upgraded acidic components of bio-oil through catalytic ketonic condensation. Energy Fuels 23(1):564–568

Ernst S, Hartmann M, Sauerbeck S, Bongers T (2000) A novel family of solid basic catalysts obtained by nitridation of crystalline microporous aluminosilicates and aluminophosphates. Appl Catal A 200(1):117–123

Fabbri D, Torri C, Baravelli V (2007) Effect of zeolites and nanopowder metal oxides on the distribution of chiral anhydrosugars evolved from pyrolysis of cellulose: an analytical study. J Anal Appl Pyrol 80(1):24–29

French R, Czernik S (2010) Catalytic pyrolysis of biomass for biofuels production. Fuel Process Technol 91(1):25–32

Gaertner CA, Serrano-Ruiz JC, Braden DJ, Dumesic JA (2009) Catalytic coupling of carboxylic acids by ketonization as a processing step in biomass conversion. J Catal 266(1):71–78

Gayubo AG, Aguayo AT, Atutxa A, Aguado R, Olazar M, Bilbao J (2004a) Transformation of oxygenate components of biomass pyrolysis oil on a HZSM-5 zeolite. II. Aldehydes, ketones, and acids. Ind Eng Chem Res 43(11):2619–2626

Gayubo AG, Aguayo AT, Atutxa A, Prieto R, Bilbao J (2004b) Deactivation of a HZSM-5 zeolite catalyst in the transformation of the aqueous fraction of biomass pyrolysis oil into hydrocarbons. Energy Fuels 18(6):1640–1647

Gayubo AG, Aguayo AT, Atutxa A, Valle B, Bilbao J (2005) Undesired components in the transformation of biomass pyrolysis oil into hydrocarbons on an HZSM-5 zeolite catalyst. J Chem Technol Biotechnol 80(11):1244–1251

Horne PA, Williams PT (1994) Premium quality fuels and chemicals from the fluidised bed pyrolysis of biomass with zeolite catalyst upgrading. Renew Energy 5(5–8):810–812

Horne PA, Williams PT (1996a) Reaction of oxygenated biomass pyrolysis model compounds over a ZSM-5 catalyst. Renew Energy 7(2):131–144

Horne PA, Williams PT (1996b) Upgrading of biomass-derived pyrolytic vapours over zeolite ZSM-5 catalyst: effect of catalyst dilution on product yields. Fuel 75(9):1043–1050

Iliopoulou E, Antonakou E, Karakoulia S, Vasalos I, Lappas A, Triantafyllidis K (2007) Catalytic conversion of biomass pyrolysis products by mesoporous materials: effect of steam stability and acidity of Al-MCM-41 catalysts. Chem Eng J 134(1):51–57

Jackson MA, Compton DL, Boateng AA (2009) Screening heterogeneous catalysts for the pyrolysis of lignin. J Anal Appl Pyrol 85(1):226–230

Lappas A, Samolada M, Iatridis D, Voutetakis S, Vasalos I (2002) Biomass pyrolysis in a circulating fluid bed reactor for the production of fuels and chemicals. Fuel 81(16):2087–2095

Lee HI, Park HJ, Park Y-K, Hur JY, Jeon J-K, Kim JM (2008) Synthesis of highly stable mesoporous aluminosilicates from commercially available zeolites and their application to the pyrolysis of woody biomass. Catal Today 132(1):68–74

Lu Q, Zhu X, Li W, Zhang Y, Chen D (2009) On-line catalytic upgrading of biomass fast pyrolysis products. Chin Sci Bull 54(11):1941–1948

Lu Q, Zhang Z-F, Dong C-Q, Zhu X-F (2010) Catalytic upgrading of biomass fast pyrolysis vapors with nano metal oxides: an analytical Py-GC/MS study. Energies 3(11):1805–1820

Mullen CA, Boateng AA (2010) Catalytic pyrolysis-GC/MS of lignin from several sources. Fuel Process Technol 91(11):1446–1458

Nilsen MH, Antonakou E, Bouzga A, Lappas A, Mathisen K, Stöcker M (2007) Investigation of the effect of metal sites in Me–Al-MCM-41 (Me = Fe, Cu or Zn) on the catalytic behavior during the pyrolysis of wooden based biomass. Microporous Mesoporous Mater 105(1):189–203

Nokkosmäki M, Kuoppala E, Leppämäki E, Krause A (2000) Catalytic conversion of biomass pyrolysis vapours with zinc oxide. J Anal Appl Pyrol 55(1):119–131

Park HJ, Dong J, Jeon J, Yoo K, Yim J, Sohn JM, Park Y (2007) Conversion of the pyrolytic vapor of radiata pine over zeolites. J Ind Eng Chem-Seoul 13(2):182

Park HJ, Heo HS, Jeon J-K, Kim J, Ryoo R, Jeong K-E, Park Y-K (2010) Highly valuable chemicals production from catalytic upgrading of radiata pine sawdust-derived pyrolytic vapors over mesoporous MFI zeolites. Appl Catal B 95(3):365–373

Pattiya A, Titiloye JO, Bridgwater AV (2008) Fast pyrolysis of cassava rhizome in the presence of catalysts. J Anal Appl Pyrol 81(1):72–79

Peralta MA, Sooknoi T, Danuthai T, Resasco DE (2009) Deoxygenation of benzaldehyde over CsNaX zeolites. J Mol Catal A Chem 312(1):78–86

Qi W, Hu C, Li G, Guo L, Yang Y, Luo J, Miao X, Du Y (2006) Catalytic pyrolysis of several kinds of bamboos over zeolite NaY. Green Chem 8(2):183–190

Samolada M, Papafotica A, Vasalos I (2000) Catalyst evaluation for catalytic biomass pyrolysis. Energy Fuels 14(6):1161–1167

Stefanidis S, Kalogiannis K, Iliopoulou E, Lappas A, Pilavachi P (2011) In-situ upgrading of biomass pyrolysis vapors: catalyst screening on a fixed bed reactor. Biores Technol 102 (17):8261–8267

Thring RW, Katikaneni SP, Bakhshi NN (2000) The production of gasoline range hydrocarbons from Alcell® lignin using HZSM-5 catalyst. Fuel Process Technol 62(1):17–30

Torri C, Reinikainen M, Lindfors C, Fabbri D, Oasmaa A, Kuoppala E (2010) Investigation on catalytic pyrolysis of pine sawdust: catalyst screening by Py-GC-MIP-AED. J Anal Appl Pyrol 88(1):7–13

Triantafyllidis KS, Iliopoulou EF, Antonakou EV, Lappas AA, Wang H, Pinnavaia TJ (2007) Hydrothermally stable mesoporous aluminosilicates (MSU-S) assembled from zeolite seeds as catalysts for biomass pyrolysis. Microporous Mesoporous Mater 99(1):132–139

Uzun BB, Sarioğlu N (2009) Rapid and catalytic pyrolysis of corn stalks. Fuel Process Technol 90 (5):705–716

Vitolo S, Bresci B, Seggiani M, Gallo M (2001) Catalytic upgrading of pyrolytic oils over HZSM-5 zeolite: behaviour of the catalyst when used in repeated upgrading–regenerating cycles. Fuel 80(1):17–26

Wang D, Xiao R, Zhang H, He G (2010) Comparison of catalytic pyrolysis of biomass with MCM-41 and CaO catalysts by using TGA–FTIR analysis. J Anal Appl Pyrol 89(2):171–177

Williams PT, Horne PA (1994) Characterisation of oils from the fluidised bed pyrolysis of biomass with zeolite catalyst upgrading. Biomass Bioenerg 7(1–6):223–236

Williams PT, Horne PA (1995a) Analysis of aromatic hydrocarbons in pyrolytic oil derived from biomass. J Anal Appl Pyrol 31:15–37

Williams PT, Horne PA (1995b) The influence of catalyst type on the composition of upgraded biomass pyrolysis oils. J Anal Appl Pyrol 31:39–61

Zhang H, Xiao R, Huang H, Xiao G (2009) Comparison of non-catalytic and catalytic fast pyrolysis of corncob in a fluidized bed reactor. Biores Technol 100(3):1428–1434

Zhu X, Lu Q, Li W, Zhang D (2010) Fast and catalytic pyrolysis of xylan: Effects of temperature and M/HZSM-5 (M = Fe, Zn) catalysts on pyrolytic products. Front Energy Power Eng Chin 4 (3):424–429

Chapter 9
Catalytic Upgrading of Bio-oil: Biomass Gasification in the Presence of Catalysts

9.1 Introduction

Because of general fact the population of the world is increasing and our ways of living, the consumption of energy was never been higher than it is today (Plouffe and Kalache 2010; Outlook 2010). Transportation sector is one of the major fields that constituting about one fifth of the total for energy consumption (Van Ruijven and van Vuuren 2009). In the future, the need for fuels will become unavoidable as the population of the world grows (Balat 2011). However, near the future this requirement constitutes become the major challenge because the crude oil which are used to produce fuels are depleting (Sorrell et al. 2010). In order to find another alternative for fuels to replace diesel and gasoline, substantial research within the energy field have been carried out. An alternative which is equivalent to the conventional fuels can be the optimal solution. Sustainable fuel which is compatible with the infrastructure, decrease the environmental man-made footprint and reduce the emission of CO_2 is required (Pachauri et al. 2014). Fuels which were derived from biomass considered to be benign for the environment and can be produced in a relatively short cycle would be the prospective fuel of the future (Balat 2011; Roedl 2010; Meinshausen et al. 2009; Demirbas 2011).

However, these technologies depend on the food grade biomass. The first generation on bio-ethanol is produced from fermentation of starch or sugar while bio-diesel is produced from fats (Venderbosch et al. 2010; Wenzel 2010). Since the efficiency of energy for required crops per unit land is relatively low and the food requirement is a constraint around the world, the utilisation of food grade biomass is a problem (McKendry 2002). For this reason, developing the production of second generation bio-fuels using other sources of biomass such as wood and agricultural waste have been the focus of new research. In order to optimize the efficiency for the paths of second generation biofuel, a lot of efforts have been spent on the route for production of liquid from biomass through syngas (Damartzis and Zabaniotou 2011; Göransson et al. 2011; Huber et al. 2006; Tijmensen et al. 2002) and also

© Springer International Publishing AG 2017
S. Bagheri, *Catalysis for Green Energy and Technology*,
Green Energy and Technology, DOI 10.1007/978-3-319-43104-8_9

synthesis of higher hydrocarbons from methanol and alcohols from syngas (Christensen et al. 2009, 2010; Keil 1999; Spath et al. 2000; Stöcker 1999). Good economy route was achieved when bio-oil was used as the platform chemical due to lower cost of transport for plants with large scale and also their flexibility with respect to the feed of biomass (Grange et al. 1996; Perego and Bosetti 2011; Rogers and Brammer 2009). Furthermore, in the current infrastructure this route also applicable. Referred as catalytic upgrading of bio-oil, the joint zeolite cracking and HDO can become routes for second-generation bio-fuels production in the future. However, both routes are still far from application in industrial scale.

9.2   Bio-oil

Both zeolite cracking and HDO are bio-oil based as platform chemical. Flash pyrolysis has been found to be a feasible route for bio-oil production and it was the most widely used (Huber et al. 2006; Elliott 2007; Elliott et al. 1990). In this section, only flash pyrolysis will be discussed and the mentioned bio-oil in the following is referred to flash pyrolysis oil (Huber et al. 2006; Akhtar and Amin 2011; Demirbaş 2000; Moffatt and Overend 1985; Peterson et al. 2008). A densification technique where both energy- and mass-density was increased when the raw material was treated at short residence time (1–2 s), high heating rates (10^3– 10^4 K/s) at intermediate temperature (300–600 °C) is called as flash pyrolysis (Bridgwater 2006). In this way, the energy density can be increased by roughly a factor of 7–8 (Badger and Fransham 2006; Raffelt et al. 2006). Virtually, any type of the biomass ranging from waste products such as chicken litter and sludge to more traditional sources such as wood and corn are compatible with the pyrolysis (Demirbas et al. 2011; Yaman 2004). Within the bio-oil, more than 300 different compounds have been identified, where the conditions and feed used in the process will influenced their specific composition (Zhang et al. 2007). With constituting about 10–30%, water is the principle species of the oil product. The oil also contains sugars, esters, carboxylic acids, furans, hydroxyaldehydes, guaiacols, hydroxyketones and phenolics (Bridgwater 1996; Goyal et al. 2008). The homogeneous appearance of the oil can be ensured with the presence of lower molecular weight oxygenated molecules, especially aldehydes and alcohols. This is because, it can act as a surfactant for the compounds of higher molecular weight that usually considered as immiscible in water and also apolar (Bridgwater et al. 2001). Overall, due to high content of water, the bio-oil has a polar nature therefore they are immiscible in the crude oil. The high oxygen and water content will results in bio-oil with low HV, which is about half of the HV of crude oil (Lu et al. 2009). The bio-oil usually have pH in the range from 2 to 4, which mainly related to the content of the formic acid and also acetic acid (Oasmaa et al. 2010). The acidic nature of the oil contents will cause a problem, because it will enhance the harsh conditions for the equipment used for transport, storage and processing. Because of corrosion, it has been proven that some common materials used for construction such as aluminium and carbon steel are not suitable when operating with bio-oil.

Instability during storage is one of the main issues with bio-oil, where it will also affect its density, HV and also viscosity. This is because of the presence of organic compounds which are highly reactive. In the presence of air, olefins are suspected to be active for repolymerization. Besides that, organic acids, aldehydes and ketones can react to form hemiacetals, acetales and ethers, respectively. The water content, viscosity and average molecular mass of oil can be increase effectively with these types of reaction. The separation of phase will ultimately occur when the quality of the oil which was seen as a function of storage time was decreased (Adjaye et al. 1992; Boucher et al. 2000; Oasmaa and Kuoppala 2003). Overall, the characteristics of the bio-oil that are not favourable is associated with the oxygenated compounds. Some of the compounds which is unfavourable were includes in the ketones, carboxylic acids and also aldehydes. Decrease in the content of oxygen is required for the oil utilisation in order to separate the organic product from water, increase the stability and also increase the HV.

9.2.1 Bio-oil Upgrading—General Considerations

Because of the high diversity of compounds in feed, the catalytic upgrading of bio-oil become a complex reaction network. For both HDO and zeolite cracking, it has been reported that hydrocracking, cracking, decarboxylation, decarbonylation, hydrogenation, polymerization and hydrodeoxygenation had taken place (Adjaye and Bakhshi 1995a, b; Wildschut et al. 2009). Besides that, in both processes formation of carbon also can be observed. Some of the examples of reactions were shown in Fig. 9.1.

The evaluation on the bio-oil was difficult because of its high diversity and also span of potential reactions and such evaluation usually restricted to model compounds. By using the data from Barin et al., the following reactions have been evaluated via thermodynamic calculations in order to get the general thermodynamic overview of the process (Barin 1997). Both Massoth et al. (2006) and Yunquan et al. (2008) have proposed a reaction path for phenol.

$$phenol + H_2 \leftrightarrow benzene + H_2O \qquad (9.1)$$

$$phenol + 4H_2 \leftrightarrow cyclohexane + H_2O \qquad (9.2)$$

When the thermodynamic equilibrium of the two reactions have been calculated, it shows that at atmospheric pressure and stoichiometric conditions with the temperatures up to at least 600 °C, the complete phenol conversion can be achieved. The thermodynamics will be shifted towards complete conversion with increasing either the excess of hydrogen or the pressure. Similar results were observed when similar calculations were made with furfural. Thus, when evaluating model

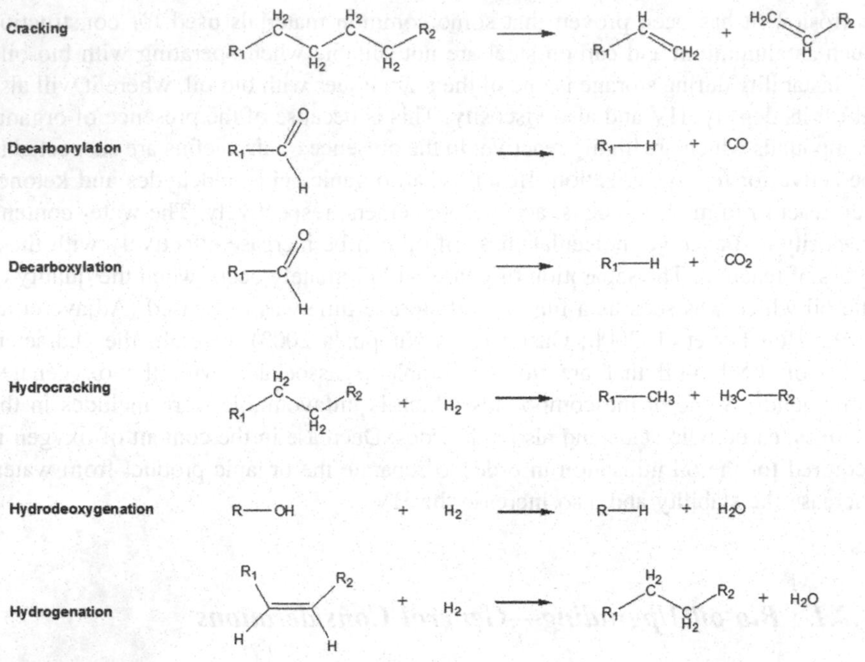

Fig. 9.1 Examples of reactions associated with catalytic bio-oil upgrading. The figure is drawn on the basis of information from Adjaye and Bakhshi (1995a), Wildschut et al. (2009)

compounds with the simplest reactions as shown in Fig. 9.1, thermodynamics not appeared to be a constraint for those processes. In practice, the conversion of each component in the bio-oil is not difficult to be evaluated. Two parameters that important are degree of deoxygenation and yield of oil:

$$Y_{oil} = \left(\frac{m_{oil}}{m_{feed}} \right) \cdot 100 \tag{9.3}$$

$$DOD = \left(1 - \frac{wt.\%_{O\,in\,product}}{wt.\%_{O\,in\,feed}} \right) \cdot 100 \tag{9.4}$$

where, Y_{oil} is the oil yields, m_{oil} is the mass of oil produced, m_{feed} is the feed mass, DOD is degree of deoxygenation and $wt\%_{o}$ is the oxygen weight percent in the oil. Together, the two important parameters can give rough overview on the extent of the reaction. The degree of deoxygenation describes the effectiveness of oxygen removal which will also indicates the quality of the oil produced, and the yield of oil can describe the selectivity to oil products. However, the parameters are less descriptive separately because in the case of no reaction, it can be seen that 100% yield can be achieved. Furthermore, the removal of some undesirable species is not related to any of the parameters and in details these should have been analysed.

9.3 Hydrodeoxygenation

Hydrodeoxygenation is closely related to hydrodesulphurization (HDS) which is a process that used in industry to eliminates sulphur from organic compounds (Chheda et al. 2007). For heteroatom exclusion, hydrogen was used by both HDS and HDO, forming H_2S and H_2O respectively. All the reactions that were shown in Fig. 9.1 are relevant for HDO. But, as the name implies, hydrodeoxygenation is the principal reaction, therefore the general overall reaction can be written as:

$$CH_{1.4}O_{0.4} + 0.7H_2 \rightarrow 1CH_2 + 0.4H_2O \qquad (9.5)$$

The "CH_2" in Eq. (9.5) represent an unspecified product of hydrocarbon. Overall, this reaction is exothermic and simple calculations have shown that when bio-oil was used, the average for heat of reaction is in the order of 2.4 MJ/kg (Catalytic 2009). In the conceptual reaction, formation of water can be observed. Therefore, at least two liquid phases which are one organic phase and one aqueous phase will be observed. Due to the production of organic compounds that have density less than water, appearance of two organic phases have been reported. In this case, a heavy oil phase will separate at below of the water and the light one on top. Usually, when the degrees of deoxygenation were high, the two organic phases will be formed instantly, and as the result, the degree of fractionation in the feed will be high.

A maximum yields of oil (56–58 wt%) have been predicted by stoichiometry of Eq. (9.5) in the case of complete deoxygenation (Bridgwater 1996). However, due to the reactions span that take place, that complete deoxygenation reaction is rarely achieved, and instead, products with residues of oxygen will be formed often. The stoichiometry of a specific experiment that have been normalized with respect to the carbon as feed was described by Venderbosch et al. (Venderbosch et al. 2010) as following:

$$CH_{1.4}O_{0.56} + 0.39H_2 \rightarrow 0.74CH_{1.4}O_{0.11} + 0.19CH_{3.02}O_{1.09} + 0.29H_2O \quad (9.6)$$

where, $CH_{1.47}O_{0.11}$ and $CH_{3.02}O_{1.09}$ is the organic aqueous phase of the product respectively. The organic phase of the hydrocarbons, some oxygen was incorporated, but when compared to pyrolysis oil (0.56), the ratios of O/C in the hydrotreated organic phase (0.11) was significantly lower. While, the ratio of O/C in the aqueous phase is higher than in the parent oil (Venderbosch et al. 2010). As reported in literature, high pressure in the range from 75 to 300 bar is commonly used (de Miguel Mercader et al. 2010; Elliott et al. 2009). Operating pressures in the range from 10 to 120 bar also have been reported by patent literature (Daudin et al. 2013; McCall et al. 2012). Better hydrogen solubility in the oil was ensured by high pressure and therefore hydrogen availability in the vicinity of the catalyst also higher. The rate of reaction will be increased and then formation of coke in the reactor will be further decreased (Venderbosch et al. 2010; Kwon et al. 2011).

When compared to the requirement of H_2 for complete deoxygenation which is around 25 mol/kg (Venderbosch et al. 2010), an excess hydrogen of 35–429 mol/kg in bio-oil have been used by Elliott et al. (2009). High residence time will favour the high degree of deoxygenation. When the LHSV was decreased from 0.70 to 0.25 h^{-1}, using a Pd/C catalyst at temperature of 340 °C and pressure of 140 bar, in a continuous flow reactor, decreased in oxygen content of the upgraded oil from 21 to 10 wt% have been shown by Elliot et al. (2009). Generally, the order of LHSV should be of 0.1–1.5 h^{-1} (McCall et al. 2012). Residence time that usually was carried out for 3–4 h is in analogy to the batch reactor tests (Gagnon and Kaliaguine 1988; Gutierrez et al. 2009). Normally, HDO was carried out at temperature of 250–450 °C (Venderbosch et al. 2010). Since the reaction is exothermic and potential full conversion of model compounds representative have been predicted by equilibrium calculations up to at least 600 °C, it appears that the operating temperature that want to be choose must be mainly based on kinetic aspects. The effects of temperature on the HDO of wood based bio-oil have been investigated by Hart and Elliot in a fixed bed reactor, at pressure of 140 bar using a Pd/C catalyst (Elliott et al. 2009). When the temperature was increased from 310 to 360 °C, the yields of oil were found to be decreased from 75 to 56%. This was also accompanied by increased in the yields of gas by a factor of 3. At 340 °C, increased in the degree of deoxygenation from 65% (at 310 °C) to 70% can be observed. No further increase in the degree of deoxygenation occur at temperature above 340 °C, but instead extensive cracking occurs. Reactivity of different types of functional groups in bio-oil have been observed by Elliott et al. (Elliott et al. 2009). On Co–MoS_2/Al_2O_3 catalyst, ketones can be deoxygenated at low temperatures (200 °C) because its activation energy for deoxygenation is relatively low. However, a significantly higher temperature is required for reaction that involved sterically hindered oxygen or more complex bound such as furans or ortho substituted phenols to proceed. By refer to this, reactivity of different compounds have been summarized as following (Furimsky 2000):

$$alcohol > ketone > alkylether > carboxylic\,acid$$
$$\approx M-/p-phenol \approx naphtol > phenol > diarylether$$
$$\approx O-phenol \approx alkylfuran > benzofuran > dibenzofuran \qquad (9.7)$$

Hydrogen consumption is one of the important aspect for the HDO reaction. In a fixed bed reactor, consumption of hydrogen in upgrading of bio-oil as a function of deoxygenation rate have been investigated by Venderbosch et al. using a Ru/C catalyst (Venderbosch et al. 2010). As a function of degree of deoxygenation, the consumption of hydrogen becomes steeply increased. It can be due to the values of the different reactivity of compounds in the bio-oil. With low consumption of hydrogen, highly reactive oxygenates such as ketones can be converted easily, but some oxygen will be bound in the compounds which are more stable. Therefore, initial saturation or hydrogenation of the molecule will accompany the more complex molecules, and thus at high degrees of deoxygenation the consumption of hydrogen will exceed the prediction of stoichiometric. Obviously, when compared to furan, the requirement of hydrogen for HDO of ketone is much lower. Overall, it means that 8 mol of hydrogen per kg of bio-oil is required in order to achieve 50% of deoxygenation.

9.3.1 Catalysts and Reaction Mechanisms

For the purpose of HDO process, many different catalysts have been tested. Different mechanisms of two different groups of catalysts, which are sulphide/oxide catalysts and transition metal catalysts will be discussed.

i. *Sulphide/oxide catalysts*

For reaction of HDO, the catalysts which were tested frequently are $Co-MoS_2$ and $Ni-MoS_2$ and these two catalysts were also being used in traditional hydrotreating process (Baldauf et al. 1994; Centeno et al. 1995; Edelman et al. 1988; Ferrari et al. 2001; French et al. 2011; Gandarias et al. 2008; Nava et al. 2009; Ryymin et al. 2010; Samolada et al. 1998; Zhang et al. 2010). In these catalysts, the electrons will be donated to the Mo atoms by Ni or Co that act as promoters. Vacancy sites of sulphur will be generated because the bond between sulphur and Mo have been weakens. In both reactions of HDO and HDS, these vacancy sites are the active sites (Badawi et al. 2009; Romero et al. 2010). The HDO of 2-ethylphenol on MoS_2-based catalysts have been studied by Romero et al. (2010) and the mechanism of the reaction was shown in Fig. 9.2. The compounds were believed to be activated by the oxygen of the molecule which adsorb the slab edge of MoS_2 vacancy site. Since the species of S–H were generated in the feed from hydrogen, this species will also be present along the catalyst's edge. The donation of proton from sulphur to the attached molecule will be enabled, thus carbocation will be formed. Direct cleavage of C–O bond can take place and form deoxygenated compound, where the removed oxygen will result in water formation.

The oxygen group that was formed from the step of deoxygenation on the metal site need to be removed as water in order for the mechanism to work. Because of catalyst transformation from sulphide to oxide form, the decreased in activity can be observed during prolonged operation. The catalyst can be stabilized and the sulphide sites can be regenerated when H_2S is co-feed to the system.

However, during the HDO of 3 wt% methyl heptanoate in m-xylene at the pressure of 15 bar and temperature of 250 °C, with $Co-MoS_2/Al_2O_3$ that was co-feed with H_2S up to 1000 ppm in a fixed bed reactor, the formation of sulphides and thiols trace amounts were shown in the study by Senol et al. (Mortensen et al. 2011). From these studies, it can be observed that when sulphide type of catalysts was used, contamination of sulphur will occur. HDS industry used $Co-MoS_2/Al_2O_3$ as catalyst where the sulphur can be removed from the oil down to a level of

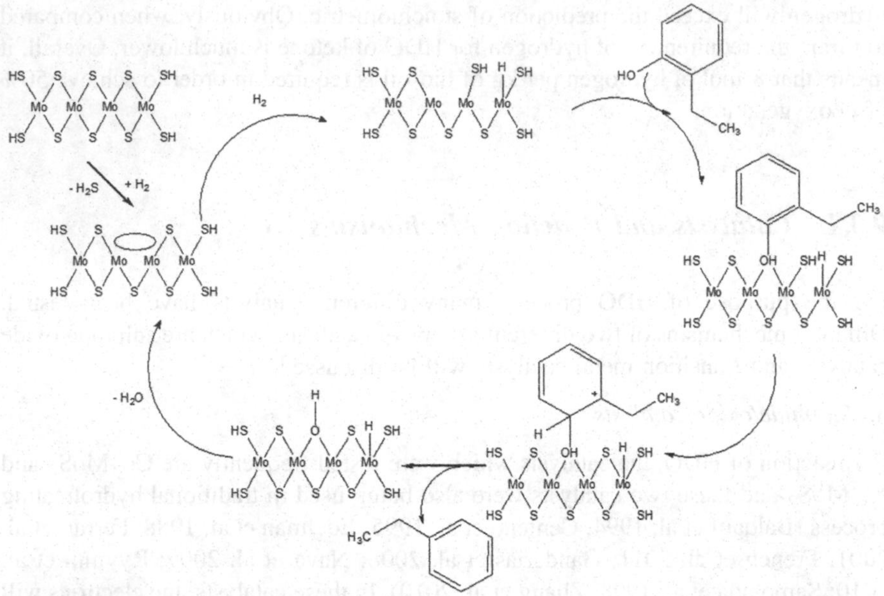

Fig. 9.2 Proposed mechanism of HDO of 2-etyhlphenol over a Co–MoS$_2$ catalyst. The circle indicates the catalytically active vacancy site. The figure is drawn on the basis of information from Romero et al. (2010)

few ppm. Besides that, when Co–MoS$_2$/C co-feed with H2S were used in order to synthesized higher alcohols from syngas, Christensen et al. show that the sulphides and thiols also were produced. Therefore, it is difficult to evaluate the effects of sulphur on this catalyst, and further attention is needed.

For the HDO with density functional theory (DFT) calculations basis, MoO$_3$ have been proposed by Moberg et al. (2010) as catalyst. As shown in Fig. 9.2, the path of deoxygenation on MoO$_3$ which are based on these calculations are similar. The presence of acid sites will affect the activity of the both sulphide and oxide type catalysts. Lewis base/acid interaction is the initial step of chemisorption, where the unsaturated metal site will be attracted by the lone pair of oxygen of the target molecule. For this reason, it was speculated that the system reactivity must partly relies on the strength and availability of the Lewis acid sites on catalyst. The concentration of relative Lewis acid site surface on different oxides were reported by Gervasini and Auroux (1991) as:

$$Cr_2O_3 > WO_3 > Nb_2O_5 > Ta_2O_5 > V_2O_5 \approx MoO_3 \qquad (9.8)$$

Against the relative Lewis acid site strength of different oxides, it should be matched. The relative strength was investigated by Dixon and Li (Li and Dixon 2006) and found as:

$$WO_3 > MoO_3 > Cr_2O_3 \qquad (9.9)$$

Donation of proton is the next step of the mechanism. It relies on the amount of hydrogen present on the catalyst, which for the oxide, the hydroxyl groups will be present. The presence of Bronsted acid hydroxyl groups on the surface of the catalyst is a must in order for the proton to have the donating capabilities. In this context, the relative Bronsted hydroxyl acidity of different oxides have been showed by Busca (Moberg et al. 2010) as:

$$WO_3 > MoO_3 > V_2O_5 > Nb_2O_5 \qquad (9.10)$$

Due to the presence of both hydroxyl sites of strong Bronsted acid and sites of strong Lewis acid, the MoO_3 was functioning as catalyst for deoxygenation reaction path. However, in a batch reactor at the temperature of 325–375 °C and pressure of 41–48 bar, HDO of 4-methylphenol by unsupported MoS_2 and MoO_3 have been investigated by Whiffen and Smith (2010), and it was found that when compared to MoS_2, MoO_3 has lower activity and higher activation energy for this reaction. Therefore, among the other catalyst of oxide type, MoO_3 might not be the best choice. Specifically, the number of acid sites available in WO_3 was higher. In a fixed bed reactor, at temperature of 150–300 °C and pressure of 115 bar, the HDO of 1 wt% phenol in n-octane have been investigated by Echeandia et al. using oxides of W and Ni–W on active carbon (Echeandia et al. 2010). For the purpose of HDO, it has been proven that these catalysts were active especially the system of Ni–W which has a potential for complete conversion of model compound. Furthermore, during the 6 h experiments, low carbon affinity can be observed. This low value suggests that the non-acidic carbon support will give beneficial effect.

ii. *Transition metal catalysts*

Transition metal catalysts also can be used to carried out the selective catalytic hydrogenation. For these systems, bifunctional catalyst is required based on the speculations made from their mechanism. Two aspects were implied by bifunctional catalyst. On one hand, the oxy-compounds must be activated, and it can be achieved by the valence of an exposed cation or an oxide form of transition metal, which often related with the support of the catalyst. Since the transition metals have the potential for hydrogen activation, this must be combined with a possibility for donation of hydrogen to the oxy-compound (Mendes et al. 2001; Stakheev and Kustov 1999; Vannice and Sen 1989; Yakovlev et al. 2009). In Fig. 9.3, the combined mechanism was shown where both activation and adsorption of oxy-compounds take place on the support.

Debates on the mechanism of hydrogenation with supported noble metal system is continuing. And generally, the donating sites of hydrogen was constituted by metals have been acknowledged, but the activation of the oxy-compound has been proposed to be either at the interface of the metal-support (Lin et al. 1994) as drawn in Fig. 9.3 or on the metal sites (Mallat and Baiker 2000; Vargas et al. 2004, 2008).

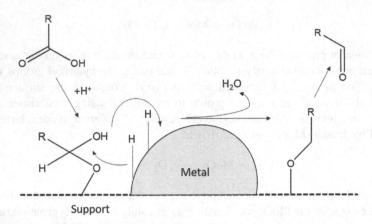

Fig. 9.3 HDO mechanism over transition metal catalysts. The mechanism drawn on the basis of information from Mendes et al. (2001), Stakheev and Kustov (1999)

It indicates that, two different paths of reaction could be used in these catalytic systems, since many of the noble metal catalysts were active for HDO.

In a batch reactor at temperature of 100 °C and pressure of 80 bar, HDO of 3 wt % guaiacol in hexadecane have been studied by Gutierrez et al. in order to investigate the activity of Pt, Pd and Rh on ZrO_2 support. The activity of the three catalyst were reported as below:

$$Rh/ZrO_2 > Co-MoS_2/Al_2O_3 > Pd/ZrO_2 > Pt/ZrO_2 \qquad (9.11)$$

To summarize, for HDO reaction, the Pd, Rh, Ru and also Pt were appeared to be the potential catalysts. However, due to their high price, they become unattractive.

In a fixed bed reactor, at atmospheric pressure and temperature at 300 °C, HDO using hydrogen or nitrogen gas that was saturated with guaiacol gases with phosphide catalyst on SiO_2 support was carried out by Zhao et al. (2011b) to measure their activity. The results based on the following relative activity was found:

$$Ni_2P/SiO_2 > Co_2P/SiO_2 > Fe_2P/SiO_2 > WP/SiO_2 > MoP/SiO_2 \qquad (9.12)$$

All the catalysts were found to be more stable than Co–MoS$_2$/Al$_2$O$_3$, and less active than Pd/Al$_2$O$_3$. Therefore, when compared to noble metal catalysts, their lower price and higher availability are the attractiveness of these catalysts. By using transition metal catalysts, different approach for HDO have been published by Zhao et al. (2009, 2010, 2011a). From this study, it has been reported that, using heterogeneous aqueous system of metal catalyst which mixed with mineral acid in a phenol/water solution, in pressure at 40 bar at temperature of 200–300 °C for 2 h, hydrogenation of phenols can be carried out. In these systems, deoxygenation can be achieved by donation of hydrogen from metal and the followed by extraction of water using mineral acid (Zhao et al. 2009, 2010). When Nafion/SiO$_2$ was combined as mineral acid, it were found that both the Raney® Ni (nickel–alumina alloy) and Pd/C can be effective catalysts (Zhao et al. 2010). However, so far, this concept was only shown in batch experiments. Besides that, the evaluation on the system potential must be carried out by study the effects of using phenol at higher concentration. Overall, it has been proved that the alternatives to both noble metal and sulphur containing type catalysts were exist, but in order to evaluate their full potential, additional development on these systems is needed.

9.3.2 Supports

In formulation of catalyst for HDO, one of the important aspect is the choice of carrier material (Yakovlev et al. 2009). Since Al$_2$O$_3$ will be converted to boemite (AlO(OH)) in the presence of higher water amounts, it is unsuitable to be use as a support (Laurent and Delmon 1994). The boemite formation will results in the nickel oxidation on the catalyst as shown by investigation on Ni–MoS$_2$/ γ-Al$_2$O$_3$ studied by Laurent and Delmon (Laurent and Delmon 1994). With respect to HDO, the nickel oxides were not active and it will block other Ni or Mo sites on the catalyst. A decrease in activity by two thirds when the catalyst was treated in a mixture of water and dodecane for 60 h was the same as decrease in activity shown when treating the catalyst with dodecane alone (Laurent and Delmon 1994). Additionally, when it was saturated in a flow of argon/phenol at temperature of 400 °C, it has been found that the species of phenolic covered the alumina by 2/3 as shown by Popov et al. (2010). In this type of support, it can be observed that high affinity for formation of carbon exists and it is believed that the surface species observed have a potential to be the carbon precursors. There was a linked between the Al$_2$O$_3$ high acidity with the high surface coverage.

Carbon was found to be a more promising support as an alternative to Al$_2$O$_3$ (Elliott et al. 2009; Maggi and Delmon 1997). When compared to Al$_2$O$_3$, carbon have an advantage in gives lower tendency for formation of carbon due to its neutral nature (Elliott et al. 2009). Similar to carbon, general neutral nature of SiO$_2$ that will results in low affinity for formation of carbon make it as a possible support for HDO (Zhao et al. 2011b). At temperature of 400 °C, in relative to the concentration that was found on Al$_2$O$_3$, only 12% of the adsorbed phenol species concentration was found on SiO$_2$ as shown by Popov et al. (2010). Interaction between the phenol and SiO$_2$ was only by hydrogen bonds, but for Al$_2$O$_3$, it can be observed that on the acid sites the surface species were adsorbed more strongly due to the dissociation of phenol (Popov et al. 2011). For the synthesis purpose, CeO$_2$ and ZrO$_2$ also have been identified as the materials with carrier potential. When compared to Al$_2$O$_3$, ZrO$_2$ have significantly less acidic character (Lin et al. 2011). As shown in Fig. 9.3, the oxy-compounds were expected to be activate on the surface of CeO$_2$ and ZrO$_2$, which then will increase the activity. Therefore, they become attractive to be formulate as new catalysts. Overall, to choose the support, two aspects must be the consideration. Firstly, they must have low affinity for formation of carbon. And secondly,

To facilitate the sufficient activity, they must have an ability to activate the oxy-compounds.

9.3.3 Deactivation

Deactivation is one of the main issues in HDO. The deactivation might occur through coking, deposition of metal, sintering of catalyst and also by water or nitrogen species. Catalyst also will affect the process of deactivation, but deposition of carbon has been proved to be the main path and general problem for deactivation of catalyst (Furimsky and Massoth 1999). Principally, formation of polyaromatic species by polycondensation and polymerization on the catalytic surface will lead to the formation of carbon. As the result, the active sites on the catalysts will be blocked (Furimsky and Massoth 1999). Specifically, due to the strong adsorption of the polyaromatic species, the builds up of carbon occurs quickly for $Co-MoS_2/Al_2O_3$. During the start-up of the system, the pore volume of the catalyst will be filled up. During initial stage for deposition of carbon, the pore volume of $Co-MoS_2/Al_2O_3$ catalyst have been filled up by one third of the total volume as reported by a study that was carried out by Fonseca et al. (1996a, 1996b), and after that since the deposition of carbon have been limited, steady state can be observed (Furimsky and Massoth 1999).

Other than feed to the system, the conditions of the process also plays an important role on the rate of the formation of carbon. When compared to saturated hydrocarbons, stronger interaction between the catalytic surface with aromatics and alkenes can be observed, which make them to have the highest affinity for formation of carbon. The conversion of hydrocarbons to carbon is more likely to happen due to their stronger binding to the surface. It also have been identified that the hydrocarbon with oxygenated groups also have high affinity for formation of carbon, through the polymerization reactions on the surface of the catalysts, especially the compounds with more than one oxygen atom (Furimsky and Massoth 1999). When the acidity of the catalysts was increased, the formation of coke also will increase since it was influenced by both Bronsted and Lewis acid sites. Lezis acid function by binding species to the surface of the catalyst. While, for Bronsted site, the protons will be donated to the relevance compounds, and then formed the carbocations which are believed to be responsible for coking (Furimsky and Massoth 1999).

Furthermore, the presence of organic acids in the feed was found to increase the affinity for formation of carbon, because the path of thermal degradation was catalysed by acids. The operating parameters must be chosen carefully in order to minimize the formation of carbon. The formation of carbon on MoS_2/Al_2O_3 can be reduced effectively by the used of hydrogen. It can be done by saturating the surface of the adsorbed species, such as alkenes and convert it to more stable molecules

(Richardson et al. 1995). The formation of carbon also was affected by the temperature of the reaction. The rate of the dehydrogenation will increase when the temperature was elevated, and it will result in the increase in rate of polycondensation. Generally, the formation of carbon will increased when the temperature of the reaction was increased (Furimsky and Massoth 1999).

9.4 Catalysts and Reaction Mechanisms

Zeolites can be observed as three-dimensional material with porous structures. To elucidate their catalytic properties and also structure, extensive work has been conducted (Stöcker 2005; Kirschhock et al. 2008; Weitkamp 2000; Xu et al. 2007). For cracking of zeolite, the mechanism is based on the series of reactions. Using general reactions of cracking, the conversion of hydrocarbons to smaller fragments will take place. The actual elimination of oxygen is associated with decarbonylation, decarboxylation and dehydration, and the main route is the dehydration reaction (Corma et al. 2007). The mechanism for zeolite dehydration of ethanol have been investigated by Bhan and Chiang as shown in Fig. 9.4 (Chiang and Bhan 2010). Adsorption on the active site will initiate the reaction. After the adsorption, two different routes have been evaluated and both routes were shown in Fig. 9.4. The two routes are either bimolecular monomer dehydration od a decomposition route.

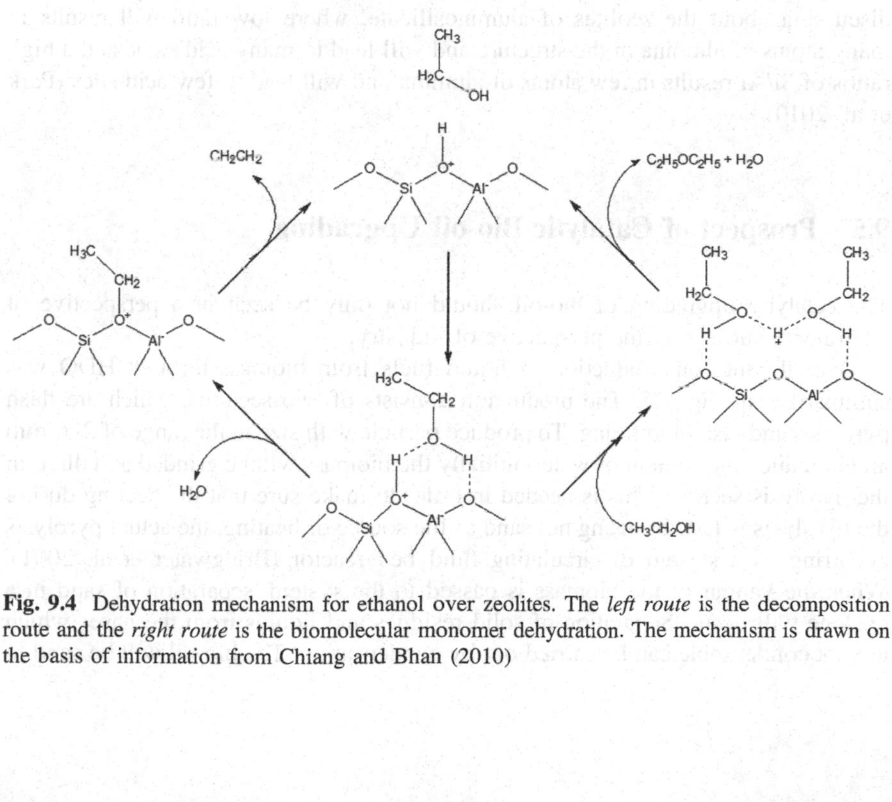

Fig. 9.4 Dehydration mechanism for ethanol over zeolites. The *left route* is the decomposition route and the *right route* is the biomolecular monomer dehydration. The mechanism is drawn on the basis of information from Chiang and Bhan (2010)

It has been concluded that the elimination of oxygen occurred by decomposition where the transition state was acted by the carbenium ion. On this basis, formation of surface ethoxide can occur, which the acid site can be regenerated and desorption can take place to form ethylene. For the dehydration of the bimolecular monomer, on the catalyst two molecules of ethanol must be presented for formation of diethylether to occur. Bhan and Chiang have been concluded that from the two routes, controlled by the zeolite pore structure is favoured, less bulky product of ethylene will be favours with small pore structures (Chiang and Bhan 2010). Therefore, it can be seen that the pore size effects the distribution of product, where production of C_9–C_{12} was increased with deoxygenation of bio-oil in larger pore size of zeolites and in medium pore size, the production of C_6–C_9 compounds was increased (Adjaye and Bakhshi 1994). Oligomerisation will accompany the decomposition reactions in the zeolite, which at the end of the reaction a mixture of larger aromatic (C_6–C_{10}) and light aliphatic (C_1–C_6) hydrocarbons were produced (Vitolo et al. 2001). Carbenium ions were formed as intermediates in the mechanism of the oligomerizing reaction (Dejaifve et al. 1980).

Therefore, in all mechanisms of relevant reaction, the carbenium ions formation is important (Huang et al. 2009; Park et al. 2010). The acid sites available is important in order to choose the catalysts. For zeolite petroleum cracking, this tendency has also been described. Where, extensive transfer of hydrogen can occur when the acid sites available is high, and it will result in high production of gasoline fraction. However, the transfer of hydrogen also affected the mechanisms of the carbon formation, therefore the fraction will be increased when many acid sites were present. The acid sites availability was related to the ratios of Si/Al when discussing about the zeolites of aluminosilicate, where low ratio will results in many atoms of alumina in the structure and will lead to many acid sites, and a high ratios of Si/Al results in few atoms of alumina and will lead to few acid sites (Park et al. 2010).

9.5 Prospect of Catalytic Bio-oil Upgrading

The catalytic upgrading of bio-oil should not only be seen in a perspective of laboratory, but also in the perspective of industry.

Overall route on production of liquid fuels from biomass through HDO was summarizes in Fig. 9.5. The production consists of two sections, which are flash pyrolysis and also biorefining. To produce particle with size in the range of 2–6 mm and to reduce the content of water, initially the biomass will be grinded and dired in the pyrolysis section. This is needed in order to make sure that the heating during the pyrolysis is fast. By using hot sand as the source of heating, the actual pyrolysis occurring as a system of circulating fluid bed reactor (Bridgwater et al. 2001). When the vapour of the biomass is passed to the system, separation of sand in a cyclone will occur. Separation of solid residues and liquids from the gases which are not condensable can be carried out by condensation. The bio-oil will be send to another site of processing or stored after the solid and oil fraction were filtered. The condenser will pass the hot off-gas to a chamber for combustion, where combustion of methane and the other potential hydrocarbons is carried out to heat up the sand for pyrolysis. The combusted off-gas which were used for biomass drying in the grinder will be used in order to achieve maximum efficiency of heat.

The production of bio-oil is preferred to be at place with smaller plants and close to the source of biomass in order to minimize the costs of transport. At the plant of biorefinery, the bio-oil will be fed into the system and will be heated at temperature of 150–280 °C and also pressurized. With respect to the catalytic reactor, treatment of thermal step in the absence of catalyst have been proposed to be incorporated with either zeolite catalyst or HDO. This should be carried out with or without the hydrogen and must take place at temperature between 200 and 300 °C. some of the compounds which are reactive in the feed can be stabilized and the reaction will be prompted and therefore in the downstream processes the affinity for formation of carbon will be lower (Bulushev and Ross 2011).

Initially, separation of heavy and light oil was carried out in the process of HDO by distillation of the oil. The processing of the fraction of heavy oil will be proceed through cracking, which here was described as FCC. Hereafter, the fraction of cracked oil will again join the fraction of light oil. Finally, separation of diesel, gasoline and etc. was carried out by distillation of the light oil. In the production of hydrogen, the utilisation of off-gasses from the FCC and HDO is required. However, production of hydrogen in required amount to be used for the synthesis in not sufficient (Jones et al. 2009). As shown in Fig. 9.5, the steam reforming has been simplified as a single step reaction, which then will be followed by the separation of hydrogen by pressure swing adsorption (PSA). The step is more complex in reality since all the pre-treatment of feed, recovery of heat and water-gas-shift were incorporated in one section (Czernik et al. 2002; Vagia and Lemonidou 2007; Wang et al. 1997, 1998).

As shown in Fig. 9.5, decrease in production of fuel by about one third of the given bio-oil amount might occurs if hydrogen is supplied from the steam reforming of the bio-oil. When the technologies are mature, it is expected that the supply of hydrogen can be done by hydrolysis with generation of energy on the basis of wind or solar energy (Agrawal and Singh 2009; Singh et al. 2010). This also will offer a path for storage of some of the solar energy. Due to the bio-oil instability, a step on stabilization potential can be inserted in between the HDO and pyrolysis. A series of parameters will decide the needs of this step. The parameters include, the time the bio-oil must be stored, the apparent stability of the specific batch of bio-oil and also the time required for transport. If no measures were taken, the bio-oil utilisation must be done in three months as indicated by works done by Oasmaa and Kuoppala. In order to increase the stability of the bio-oil, different methods have been suggested. The reactivity must be decreased when the bio-oil was mixed with alcohols (Triantafillidis et al. 2001). Furthermore, thermal hydrotreatment at low temperature (100–200 °C) has been proposed since it will prompt the cracking and hydrodeoxygenation of some of the groups which are reactive (Grange et al. 1996). Overall, it can be concluded that before the HDO can

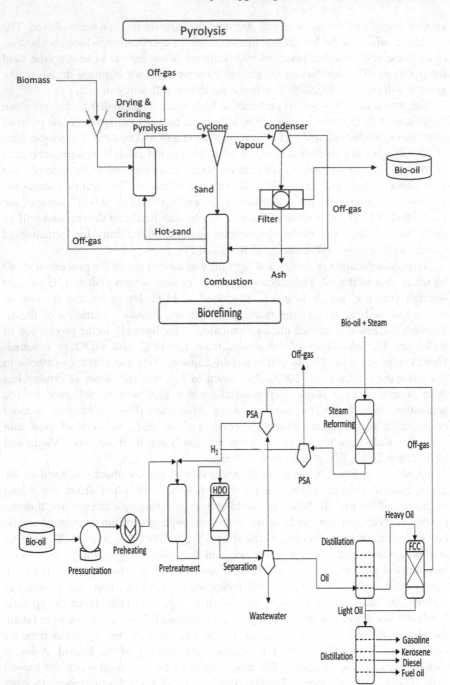

Fig. 9.5 Overall flow sheet for the production of bio-fuels on the basis of catalytic upgrading of bio-oil. The figure is based on the information from Jones et al (2009)

be used in industrial scale, it must be tested in a series of fields. Therefore, in the future, some tasks are required to be fulfilled. The tasks are as follows:

i. To direct the effort, development of catalyst and new formulations must be investigated.
ii. The understanding on mechanism for formation of carbon from different classes of compounds must be improved.
iii. Understanding on the kinetics of HDO of bio-oil and model compounds also must be improved.
iv. The effects of impurities on the performance of bio-oil using different catalysts must be further studied.
v. Better understanding on decrease of reaction temperature and partial pressure of hydrogen.
vi. Defining the requirement for the degree of oxygen removal in the context of further refining.
vii. Finding (sustainable) sources for hydrogen.

References

Adjaye JD, Bakhshi N (1994) Upgrading of a wood-derived oil over various catalysts. Biomass Bioenerg 7(1):201–211
Adjaye JD, Bakhshi N (1995a) Catalytic conversion of a biomass-derived oil to fuels and chemicals I: model compound studies and reaction pathways. Biomass Bioenerg 8(3):131–149
Adjaye JD, Bakhshi N (1995b) Production of hydrocarbons by catalytic upgrading of a fast pyrolysis bio-oil. Part I: Conversion over various catalysts. Fuel Process Technol 45(3): 161–183
Adjaye JD, Sharma RK, Bakhshi NN (1992) Characterization and stability analysis of wood-derived bio-oil. Fuel Process Technol 31(3):241–256
Agrawal R, Singh NR (2009) Synergistic routes to liquid fuel for a petroleum-deprived future. AIChE J 55(7):1898–1905
Akhtar J, Amin NAS (2011) A review on process conditions for optimum bio-oil yield in hydrothermal liquefaction of biomass. Renew Sustain Energy Rev 15(3):1615–1624
Badawi M, Cristol S, Paul J-F, Payen E (2009) DFT study of furan adsorption over stable molybdenum sulfide catalyst under HDO conditions. C R Chim 12(6):754–761
Badger PC, Fransham P (2006) Use of mobile fast pyrolysis plants to densify biomass and reduce biomass handling costs—A preliminary assessment. Biomass Bioenerg 30(4):321–325
Balat M (2011) Production of bioethanol from lignocellulosic materials via the biochemical pathway: a review. Energy Convers Manag 52(2):858–875
Baldauf W, Balfanz U, Rupp M (1994) Upgrading of flash pyrolysis oil and utilization in refineries. Biomass Bioenerg 7(1–6):237–244
Barin I (1997) Thermochemical data of pure substances, thermochemical data of pure substances. Wiley-VCH, Germany
Boucher M, Chaala A, Roy C (2000) Bio-oils obtained by vacuum pyrolysis of softwood bark as a liquid fuel for gas turbines. Part I: properties of bio-oil and its blends with methanol and a pyrolytic aqueous phase. Biomass Bioenerg 19(5):337–350
Bridgwater A (1996) Production of high grade fuels and chemicals from catalytic pyrolysis of biomass. Catal Today 29(1–4):285–295

Bridgwater T (2006) Biomass for energy. J Sci Food Agric 86(12):1755–1768

Bridgwater A, Czernik S, Piskorz J (2001) An overview of fast pyrolysis. In: Progress in thermochemical biomass conversion, pp 977–997

Bulushev DA, Ross JR (2011) Catalysis for conversion of biomass to fuels via pyrolysis and gasification: a review. Catal Today 171(1):1–13

by Catalytic TF (2009) Pyrolysis oil upgrading to transportation fuels by catalytic hydrotreatment

Centeno A, Laurent E, Delmon B (1995) Influence of the support of CoMo sulfide catalysts and of the addition of potassium and platinum on the catalytic performances for the hydrodeoxygenation of carbonyl, carboxyl, and guaiacol-type molecules. J Catal 154(2):288–298

Chheda JN, Huber GW, Dumesic JA (2007) Liquid-phase catalytic processing of biomass-derived oxygenated hydrocarbons to fuels and chemicals. Angew Chem Int Ed 46(38):7164–7183

Chiang H, Bhan A (2010) Catalytic consequences of hydroxyl group location on the rate and mechanism of parallel dehydration reactions of ethanol over acidic zeolites. J Catal 271 (2):251–261

Christensen JM, Mortensen PM, Trane R, Jensen PA, Jensen AD (2009) Effects of H_2S and process conditions in the synthesis of mixed alcohols from syngas over alkali promoted cobalt-molybdenum sulfide. Appl Catal A 366(1):29–43

Christensen JM, Jensen PA, Schiødt NC, Jensen AD (2010) Coupling of Alcohols over alkali-promoted cobalt–molybdenum sulfide. ChemCatChem 2(5):523–526

Corma A, Huber GW, Sauvanaud L, O'connor P (2007) Processing biomass-derived oxygenates in the oil refinery: catalytic cracking (FCC) reaction pathways and role of catalyst. J Catal 247 (2):307–327

Czernik S, French R, Feik C, Chornet E (2002) Hydrogen by catalytic steam reforming of liquid byproducts from biomass thermoconversion processes. Ind Eng Chem Res 41(17):4209–4215

Damartzis T, Zabaniotou A (2011) Thermochemical conversion of biomass to second generation biofuels through integrated process design—A review. Renew Sustain Energy Rev 15(1): 366–378

Daudin A, Bournay L, Chapus T (2013) Method of converting effluents of renewable origin into fuel of excellent quality by using a molybdenum-based catalyst. Google Patents

de Miguel MF, Groeneveld M, Kersten S, Way N, Schaverien C, Hogendoorn J (2010) Production of advanced biofuels: co-processing of upgraded pyrolysis oil in standard refinery units. Appl Catal B 96(1):57–66

Dejaifve P, Védrine JC, Bolis V, Derouane EG (1980) Reaction pathways for the conversion of methanol and olefins on H-ZSM-5 zeolite. J Catal 63(2):331–345

Demirbas A (2011) Competitive liquid biofuels from biomass. Appl Energy 88(1):17–28

Demirbaş A (2000) Mechanisms of liquefaction and pyrolysis reactions of biomass. Energy Convers Manag 41(6):633–646

Demirbas MF, Balat M, Balat H (2011) Biowastes-to-biofuels. Energy Convers Manag 52 (4):1815–1828

Echeandia S, Arias P, Barrio V, Pawelec B, Fierro J (2010) Synergy effect in the HDO of phenol over Ni–W catalysts supported on active carbon: Effect of tungsten precursors. Appl Catal B 101(1):1–12

Edelman MC, Maholland MK, Baldwin RM, Cowley SW (1988) Vapor-phase catalytic hydrodeoxygenation of benzofuran. J Catal 111(2):243–253

Elliott DC (2007) Historical developments in hydroprocessing bio-oils. Energy Fuels 21(3):1792–1815

Elliott D, Baker E, Beckman D, Solantausta Y, Tolenhiemo V, Gevert S, Hörnell C, Östman A, Kjellström B (1990) Technoeconomic assessment of direct biomass liquefaction to transportation fuels. Biomass 22(1–4):251–269

Elliott DC, Hart TR, Neuenschwander GG, Rotness LJ, Zacher AH (2009) Catalytic hydroprocessing of biomass fast pyrolysis bio-oil to produce hydrocarbon products. Environ Prog Sustain Energy 28(3):441–449

Ferrari M, Bosmans S, Maggi R, Delmon B, Grange P (2001) CoMo/carbon hydrodeoxygenation catalysts: influence of the hydrogen sulfide partial pressure and of the sulfidation temperature. Catal Today 65(2):257–264

Fonseca A, Zeuthen P, Nagy JB (1996a) 13C nmr quantitative analysis of catalyst carbon deposits. Fuel 75(12):1363–1376

Fonseca A, Zeuthen P, Nagy JB (1996b) Assignment of an average chemical structure to catalyst carbon deposits on the basis of quantitative 13C nmr spectra. Fuel 75(12):1413–1423

French RJ, Stunkel J, Baldwin RM (2011) Mild hydrotreating of bio-oil: effect of reaction severity and fate of oxygenated species. Energy Fuels 25(7):3266–3274

Furimsky E (2000) Catalytic hydrodeoxygenation. Appl Catal A 199(2):147–190

Furimsky E, Massoth FE (1999) Deactivation of hydroprocessing catalysts. Catal Today 52 (4):381–495

Gagnon J, Kaliaguine S (1988) Catalytic hydrotreatment of vacuum pyrolysis oils from wood. Ind Eng Chem Res 27(10):1783–1788

Gandarias I, Barrio V, Requies J, Arias P, Cambra J, Güemez M (2008) From biomass to fuels: Hydrotreating of oxygenated compounds. Int J Hydrogen Energy 33(13):3485–3488

Gervasini A, Auroux A (1991) Acidity and basicity of metal oxide surfaces II. Determination by catalytic decomposition of isopropanol. J Catal 131(1):190–198

Göransson K, Söderlind U, He J, Zhang W (2011) Review of syngas production via biomass DFBGs. Renew Sustain Energy Rev 15(1):482–492

Goyal H, Seal D, Saxena R (2008) Bio-fuels from thermochemical conversion of renewable resources: a review. Renew Sustain Energy Rev 12(2):504–517

Grange P, Laurent E, Maggi R, Centeno A, Delmon B (1996) Hydrotreatment of pyrolysis oils from biomass: reactivity of the various categories of oxygenated compounds and preliminary techno-economical study. Catal Today 29(1–4):297–301

Gutierrez A, Kaila R, Honkela M, Slioor R, Krause A (2009) Hydrodeoxygenation of guaiacol on noble metal catalysts. Catal Today 147(3):239–246

Huang J, Long W, Agrawal PK, Jones CW (2009) Effects of acidity on the conversion of the model bio-oil ketone cyclopentanone on H–Y zeolites. J Phys Chem C 113(38):16702–16710

Huber GW, Iborra S, Corma A (2006) Synthesis of transportation fuels from biomass: chemistry, catalysts, and engineering. Chem Rev 106(9):4044–4098

Jones SB, Valkenburg C, Walton CW, Elliott DC, Holladay JE, Stevens DJ, Kinchin C, Czernik S (2009) Production of gasoline and diesel from biomass via fast pyrolysis, hydrotreating and hydrocracking: a design case. Pacific Northwest National Laboratory, Richland, WA

Keil FJ (1999) Methanol-to-hydrocarbons: process technology. Microporous Mesoporous Mater 29(1):49–66

Kirschhock CE, Feijen EJ, Jacobs PA, Martens JA (2008) Hydrothermal zeolite synthesis. In: Handbook of heterogeneous catalysis

Kwon KC, Mayfield H, Marolla T, Nichols B, Mashburn M (2011) Catalytic deoxygenation of liquid biomass for hydrocarbon fuels. Renew Energy 36(3):907–915

Laurent E, Delmon B (1994) Influence of water in the deactivation of a sulfided NiMoγ-Al₂O₃ catalyst during hydrodeoxygenation. J Catal 146(1):281288–285291

Li S, Dixon DA (2006) Molecular and electronic structures, Brönsted basicities, and Lewis acidities of group VIB transition metal oxide clusters. J Phys Chem A 110(19):6231–6244

Lin SD, Sanders DK, Vannice MA (1994) Influence of metal-support effects on acetophenone hydrogenation over platinum. Appl Catal A 113(1):59–73

Lin Y-C, Li C-L, Wan H-P, Lee H-T, Liu C-F (2011) Catalytic hydrodeoxygenation of guaiacol on Rh-based and sulfided CoMo and NiMo catalysts. Energy Fuels 25(3):890–896

Lu Q, Li W-Z, Zhu X-F (2009) Overview of fuel properties of biomass fast pyrolysis oils. Energy Convers Manag 50(5):1376–1383

Maggi R, Delmon B (1997) A review of catalytic hydrotreating processes for the upgrading of liquids produced by flash pyrolysis. Stud Surf Sci Catal 106:99–113

Mallat T, Baiker A (2000) Selectivity enhancement in heterogeneous catalysis induced by reaction modifiers. Appl Catal A 200(1):3–22

Massoth F, Politzer P, Concha M, Murray J, Jakowski J, Simons J (2006) Catalytic hydrodeoxygenation of methyl-substituted phenols: correlations of kinetic parameters with molecular properties. J Phys Chem B 110(29):14283–14291

McCall MJ, Brandvold TA, Elliott DC (2012) Fuel and fuel blending components from biomass derived pyrolysis oil. Google Patents

McKendry P (2002) Energy production from biomass (part 1): overview of biomass. Biores Technol 83(1):37–46

Meinshausen M, Meinshausen N, Hare W, Raper SC, Frieler K, Knutti R, Frame DJ, Allen MR (2009) Greenhouse-gas emission targets for limiting global warming to 2 C. Nature 458 (7242):1158–1162

Mendes M, Santos O, Jordao E, Silva A (2001) Hydrogenation of oleic acid over ruthenium catalysts. Appl Catal A 217(1):253–262

Moberg DR, Thibodeau TJ, Amar FG, Frederick BG (2010) Mechanism of hydrodeoxygenation of acrolein on a cluster model of MoO3. J Phys Chem C 114(32):13782–13795

Moffatt J, Overend R (1985) Direct liquefaction of wood through solvolysis and catalytic hydrodeoxygenation: an engineering assessment. Biomass 7(2):99–123

Mortensen PM, Grunwaldt J-D, Jensen PA, Knudsen K, Jensen AD (2011) A review of catalytic upgrading of bio-oil to engine fuels. Appl Catal A 407(1):1–19

Nava R, Pawelec B, Castaño P, Álvarez-Galván M, Loricera C, Fierro J (2009) Upgrading of bio-liquids on different mesoporous silica-supported CoMo catalysts. Appl Catal B 92(1): 154–167

Oasmaa A, Kuoppala E (2003) Fast pyrolysis of forestry residue. 3. Storage stability of liquid fuel. Energy Fuels 17(4):1075–1084

Oasmaa A, Elliott DC, Korhonen J (2010) Acidity of biomass fast pyrolysis bio-oils. Energy Fuels 24(12):6548–6554

Outlook AE (2010) Energy information administration. Department of Energy 2010 (9)

Pachauri RK, Allen MR, Barros VR, Broome J, Cramer W, Christ R, Church JA, Clarke L, Dahe Q, Dasgupta P (2014) Climate change 2014: synthesis report. In: Contribution of working groups I, II and III to the fifth assessment report of the intergovernmental panel on climate change. IPCC

Park HJ, Heo HS, Jeon J-K, Kim J, Ryoo R, Jeong K-E, Park Y-K (2010) Highly valuable chemicals production from catalytic upgrading of radiata pine sawdust-derived pyrolytic vapors over mesoporous MFI zeolites. Appl Catal B 95(3):365–373

Perego C, Bosetti A (2011) Biomass to fuels: the role of zeolite and mesoporous materials. Microporous Mesoporous Mater 144(1):28–39

Peterson AA, Vogel F, Lachance RP, Fröling M, Antal MJ Jr, Tester JW (2008) Thermochemical biofuel production in hydrothermal media: a review of sub-and supercritical water technologies. Energy Environ Sci 1(1):32–65

Plouffe L, Kalache A (2010) Towards global age-friendly cities: determining urban features that promote active aging. J Urban Health 87(5):733–739

Popov A, Kondratieva E, Goupil JM, Mariey L, Bazin P, Gilson J-P, Travert A, Maugé F (2010) Bio-oils hydrodeoxygenation: adsorption of phenolic molecules on oxidic catalyst supports. J Phys Chem C 114(37):15661–15670

Popov A, Kondratieva E, Gilson J-P, Mariey L, Travert A, Maugé F (2011) IR study of the interaction of phenol with oxides and sulfided CoMo catalysts for bio-fuel hydrodeoxygenation. Catal Today 172(1):132–135

Raffelt K, Henrich E, Koegel A, Stahl R, Steinhardt J, Weirich F (2006) The BTL2 process of biomass utilization entrained-flow gasification of pyrolyzed biomass slurries. Appl Biochem Biotechnol 129(1–3):153–164

Richardson S, Nagaishi H, Gray M (1995) Initial carbon deposition on a NiMo/Gamma-Al$_2$O$_3$ bitumen hydrocracking catalyst: the effect of reaction time and hydrogen pressure. Prepr Am Chem Soc Div Pet Chem 40(3):455–459

Roedl A (2010) Production and energetic utilization of wood from short rotation coppice—a life cycle assessment. Int J Life Cycle Assess 15(6):567–578

Rogers J, Brammer JG (2009) Analysis of transport costs for energy crops for use in biomass pyrolysis plant networks. Biomass Bioenerg 33(10):1367–1375

Romero Y, Richard F, Brunet S (2010) Hydrodeoxygenation of 2-ethylphenol as a model compound of bio-crude over sulfided Mo-based catalysts: promoting effect and reaction mechanism. Appl Catal B 98(3):213–223

Ryymin E-M, Honkela ML, Viljava T-R, Krause AOI (2010) Competitive reactions and mechanisms in the simultaneous HDO of phenol and methyl heptanoate over sulphided NiMo/γ-Al₂O₃. Appl Catal A 389(1):114–121

Samolada M, Baldauf W, Vasalos I (1998) Production of a bio-gasoline by upgrading biomass flash pyrolysis liquids via hydrogen processing and catalytic cracking. Fuel 77(14):1667–1675

Singh NR, Delgass WN, Ribeiro FH, Agrawal R (2010) Estimation of liquid fuel yields from biomass. Environ Sci Technol 44(13):5298–5305

Sorrell S, Speirs J, Bentley R, Brandt A, Miller R (2010) Global oil depletion: a review of the evidence. Energy Policy 38(9):5290–5295

Spath PL, Lane JM, Mann MK, Amos W (2000) Update of hydrogen from biomass-determination of the delivered cost of hydrogen. Milestone report for the US Department of Energy's hydrogen program

Stakheev AY, Kustov L (1999) Effects of the support on the morphology and electronic properties of supported metal clusters: modern concepts and progress in 1990s. Appl Catal A 188(1):3–35

Stöcker M (1999) Methanol-to-hydrocarbons: catalytic materials and their behavior. Microporous Mesoporous Mater 29(1):3–48

Stöcker M (2005) Gas phase catalysis by zeolites. Microporous Mesoporous Mater 82(3):257–292

Tijmensen MJ, Faaij AP, Hamelinck CN, van Hardeveld MR (2002) Exploration of the possibilities for production of Fischer Tropsch liquids and power via biomass gasification. Biomass Bioenerg 23(2):129–152

Triantafillidis CS, Vlessidis AG, Nalbandian L, Evmiridis NP (2001) Effect of the degree and type of the dealumination method on the structural, compositional and acidic characteristics of H-ZSM-5 zeolites. Microporous Mesoporous Mater 47(2):369–388

Vagia EC, Lemonidou AA (2007) Thermodynamic analysis of hydrogen production via steam reforming of selected components of aqueous bio-oil fraction. Int J Hydrogen Energy 32 (2):212–223

Van Ruijven B, van Vuuren DP (2009) Oil and natural gas prices and greenhouse gas emission mitigation. Energy Policy 37(11):4797–4808

Vannice MA, Sen B (1989) Metal-support effects on the intramolecular selectivity of crotonaldehyde hydrogenation over platinum. J Catal 115(1):65–78

Vargas A, Bürgi T, Baiker A (2004) Adsorption of activated ketones on platinum and their reactivity to hydrogenation: a DFT study. J Catal 222(2):439–449

Vargas A, Reimann S, Diezi S, Mallat T, Baiker A (2008) Adsorption modes of aromatic ketones on platinum and their reactivity towards hydrogenation. J Mol Catal A Chem 282(1):1–8

Venderbosch R, Ardiyanti A, Wildschut J, Oasmaa A, Heeres H (2010) Stabilization of biomass-derived pyrolysis oils. J Chem Technol Biotechnol 85(5):674–686

Vitolo S, Bresci B, Seggiani M, Gallo M (2001) Catalytic upgrading of pyrolytic oils over HZSM-5 zeolite: behaviour of the catalyst when used in repeated upgrading–regenerating cycles. Fuel 80(1):17–26

Wang D, Czernik S, Montane D, Mann M, Chornet E (1997) Biomass to hydrogen via fast pyrolysis and catalytic steam reforming of the pyrolysis oil or its fractions. Ind Eng Chem Res 36(5):1507–1518

Wang D, Czernik S, Chornet E (1998) Production of hydrogen from biomass by catalytic steam reforming of fast pyrolysis oils. Energy Fuels 12(1):19–24

Weitkamp J (2000) Zeolites and catalysis. Solid State Ionics 131(1):175–188

Wenzel H (2010) Breaking the biomass bottleneck of the fossil free society. Concito

Whiffen VM, Smith KJ (2010) Hydrodeoxygenation of 4-methylphenol over unsupported MoP, MoS₂, and MoOₓ catalysts. Energy Fuels 24(9):4728–4737

Wildschut J, Mahfud FH, Venderbosch RH, Heeres HJ (2009) Hydrotreatment of fast pyrolysis oil using heterogeneous noble-metal catalysts. Ind Eng Chem Res 48(23):10324–10334

Xu R, Pang W, Yu J, Huo Q, Chen J (2007) Structural chemistry of microporous materials, chemistry of zeolites and related porous materials. John Wiley & Sons (Asia) Pte Ltd., Singapore

Yakovlev V, Khromova S, Sherstyuk O, Dundich V, Ermakov DY, Novopashina V, Lebedev MY, Bulavchenko O, Parmon V (2009) Development of new catalytic systems for upgraded bio-fuels production from bio-crude-oil and biodiesel. Catal Today 144(3):362–366

Yaman S (2004) Pyrolysis of biomass to produce fuels and chemical feedstocks. Energy Convers Manag 45(5):651–671

Yunquan Y, Gangsheng T, Smith KJ, Tye CT (2008) Hydrodeoxygenation of phenolic model compounds over MoS$_2$ catalysts with different structures. Chin J Chem Eng 16(5):733–739

Zhang Q, Chang J, Wang T, Xu Y (2007) Review of biomass pyrolysis oil properties and upgrading research. Energy Convers Manag 48(1):87–92

Zhang W, Zhang Y, Zhao L, Wei W (2010) Catalytic activities of NiMo carbide supported on SiO$_2$ for the hydrodeoxygenation of ethyl benzoate, acetone, and acetaldehyde. Energy Fuels 24 (3):2052–2059

Zhao C, Kou Y, Lemonidou AA, Li X, Lercher JA (2009) Highly selective catalytic conversion of phenolic bio-oil to alkanes. Angew Chem 121(22):4047–4050

Zhao C, Kou Y, Lemonidou AA, Li X, Lercher JA (2010) Hydrodeoxygenation of bio-derived phenols to hydrocarbons using RANEY® Ni and Nafion/SiO$_2$ catalysts. Chem Commun 46 (3):412–414

Zhao C, He J, Lemonidou AA, Li X, Lercher JA (2011a) Aqueous-phase hydrodeoxygenation of bio-derived phenols to cycloalkanes. J Catal 280(1):8–16

Zhao H, Li D, Bui P, Oyama S (2011b) Hydrodeoxygenation of guaiacol as model compound for pyrolysis oil on transition metal phosphide hydroprocessing catalysts. Appl Catal A 391 (1):305–310

Chapter 10
Production of Renewable Hydrogen; *Liquid* Transportation Fuels (*BTL*)

10.1 Introduction

The main challenges in transportation sector at United States are the volatility of the global oil market, the prices of energy highly increasing and consistent pressure to reduce the emissions of the lifecycle greenhouse gas. Due to the concern on the "peak-oil" domestic for production of crude oil, the nations of OPEC are remains unrest (Wang et al. 2013; Nashawi et al. 2010), and efforts to develop liquid fuels which can be produced from domestic feedstocks based on carbon have been motivated in order to provide a significant means for increasing security of nation by enhanced the independence of energy. Therefore, it is expected that the demand for additional fuel using "non-petroleum" based feedstocks will be largely satisfied for the country. To process the alternative feedstocks, investigation on several technologies have been carried out. The main supply of the liquid fuels come from the biomass sources.

Biomass is a renewable energy source. During photosynthesis, the atmospheric CO_2 can be absorbed by biomass (Council 2008; Laser et al. 2009b; Council 2010), therefore it was able to solve the concerns on both enhancing production of domestic fuels and also reduction of the emissions of lifecycle greenhouse gas. The future of energy in United States can be shaped if the biomass were harvested sustainably. However, for production of liquid fuels, the land used to grow the biomass must be analysed in the context of potential requirements of the land such as feed, food and also natural habitats preservation. Ultimately, the investigation on removal of forest and crop residues for production of biofuel must be carried out using a framework of holistic. Connection between multiple processes which occur on both ecosystem and the farm such as mitigation of erosion, management of soil carbon, management of nutrient, quality of water or air and production of feed or food can be recognised by this framework.

Nevertheless, without any deforestation and disruption of animal and human food, significant supply of biomass feedstocks can also be generated in a

© Springer International Publishing AG 2017
S. Bagheri, *Catalysis for Green Energy and Technology*,
Green Energy and Technology, DOI 10.1007/978-3-319-43104-8_10

sustainable manner (Perlack et al. 2005). In fact, currently about 400–500 million tons of biomass are available for production of biofuel and in the future, it is estimated to be about 1.3 billion tons biomass available (Council 2010; Perlack et al. 2005). When compared to liquid fuels derived from biomass, the same environmental benefit will not be achieved by liquid fuels that were derived from natural gas or coal. Due to the low delivered cost of coal when compared to natural gas and also biomass, the interest in the plant with coal based has been continued (Council 2010; Larson et al. 2009; Kreutz et al. 2008). However, the conversion of significant portion of carbon feedstock to CO_2 might be required due to the high content of carbon in the coal (Agrawal et al. 2007; Baliban et al. 2010, 2011, 2012; Elia et al. 2011, 2012). The conversion rates of carbon feedstock to final liquid fuels can be increased by high ratio of hydrogen to carbon which can be provided by natural gas. The delivered cost of natural gas can be reduced by recent prospects for production of shale gas. This make the natural gas is more attractive to be used as feedstock for production of liquid fuels.

When compared to single feedstock refineries, hybrid refineries that used both biomass and fossil-based feedstock are more environmentally and economically superior because it takes advantage of strength of each feed. The hybrid refineries with utilisation of biomass can be classified into three, which are liquids from biomass/coal (Kreutz et al. 2008; Larson et al. 2010; Chen et al. 2011; Liu et al. 2010, 2011b; Warren and El-Halwagi 1996; Williams et al. 2011), liquids from biomass/natural gas (Liu et al. 2011a; Borgwardt 1997; Li et al. 2010; Dong and Steinberg 1997), and liquids from natural gas/biomass/coal (Baliban et al. 2010, 2011; Onel et al. 2016). In United States, to prevent strong dependence to the supplies of petroleum, it is important for the national scale to use the natural gas, biomass and also coal as an economic and environmentally suitable supplies. In order to introduce alternative design of process for production of diesel, kerosene and also gasoline using any one or a combination of natural gas, biomass or coal feedstocks, the development key have been made by research groups worldwide as highlighted in recent review (Floudas et al. 2012).

Even though the benefits advantage of using fossil-based feedstocks which are low-cost can be taken by the hybrid-feedstock refineries, the use of multiple feedstocks in single refinery is not always practical. The locations of biomass feedstocks in its distributed network will consist of few points where infrastructure for delivery of natural gas or coal is not exist or minimal. It is very important to harness the potential of this resource in an economic and sustainable ways in order to develop the system of biomass to liquid fuel (BTL). The thermochemical-based systems of BTL is high cost in both their feedstock of biomass and investment of capital, therefore there is strong desire to develop the novel processes which can use the unit operations that already exist in a topological design that is more efficient.

In order to analyse the design of one particular type of process, generally, the topology will be fixed based on the previous studies of BTL processes. The determination of the plant financial metrics is carried out by analysis of economic and also calculations of mass and heat balances of the process using software of standard simulation (Laser et al. 2009a; Tock et al. 2010; Hamelinck et al. 2004;

Perales et al. 2011; Tijmensen et al. 2002; Bridgwater and Double 1991; Wright et al. 2010; Clausen et al. 2010; Sharma et al. 2011; Henrich et al. 2009; Sunde et al. 2011). Significant amount of manpower and computational time were required in order to determine the design which are "best possible", such as the process design of plethora as reported in academic literature. A framework of standard set of environmental targets, product requirements, financial parameters, feedstock properties and conditions of unit operating for all designs were required to fairly and thoroughly compare the design of the processes. In order to consider and become intractable over some possibilities of upper threshold, the effort to conduct such scales of analysis were required.

Moreover, there is no guarantee that this strategy is the best design when compared to topologies of novel process that were not considered in the comparative analysis. The framework of the process synthesis will be used to examine (i) gasification of biomass with or without the recycle of syngas, (ii) conversion of syngas via synthesis of methanol of Fischer-Tropsch (FT), (iii) conversion of methanol via methanol-to-olefins (MTO) or methanol-to-gasoline (MTG) and (iv) upgrading of hydrocarbon via oligomerization of olefin or fractionation of carbon number, ZSM-5 zeolite catalysis and subsequent treatment. The key products from the refinery of BTL will be diesel, kerosene and gasoline with allowable by products of electricity and liquefied petroleum gas (LPG). The capability of the framework of process synthesis were demonstrated using selected case studies as illustrated by the quantitative trade-offs related with key metrics for the refinery of BTL.

10.2 Catalytic Production of Liquid Hydrocarbon Transportation Fuels

The route of biomass to liquids (BTL) similar to gas to liquids (GTL) or coal to liquids (CTL) technologies. It refers to conversion of liquid hydrocarbon fuels from biomass by integration of two different processes: hydrocarbon fuels from Fischer-Tropsch synthesis (FTS) and syngas from gasification of biomass. Individually, development of both technologies is relatively well. In the early 1900s, the first industrial process which is FTS and gasification of biomass (resembles gasification of coal) was developed, and has been used extensively for production of liquid hydrocarbon fuels in countries like South Africa. When requirement of efficient integration for both technologies are required, the challenges for BTL are arise. Therefore, for syngas production, new issues and challenges were introduced when the classical feeds such as natural gas and coal were replaced by using biomass. Thermal degradation in the presence of oxygen-containing agent such as oxygen, air and steam is described as gasification. In order to generate mixtures which are rich in H_2 and CO (syngas) or for production of gaseous stream with high heating value which composed of H_2, CO,

CH_4, CO_2 and N_2 (producer gas), the combustion of the biomass must be partially done by controlling the atmosphere of the reaction (Bridgwater et al. 2001).

In production of syngas streams, these streams are considered as the feedstocks for chemicals and fuels, and pure oxygen is utilised as the oxidizing agent, while for production of producer gas, air is utilised as the oxidizing agent and later will be combusted to produce heat and also electricity. High temperatures will favour the gasification of biomass to syngas because endothermic reaction was involved in the decomposition of carbohydrates to syngas (Lange 2007). However, it is difficult to control the composition of the gas from the gasifier when the reaction was carried out at such harsh conditions. The composition of the gas will depend on several factors such as particle size, source of biomass, design of the gasifier and also the conditions of the gasification. As mentioned above, by favouring the partial oxidation reactions, mixtures which were rich in syngas can be achieved when the oxygen co-feeding was below the regime of the stoichiometric (Lange 2007). Both composition of gas stream and rate of gasification were affected by the particle size of the biomass. In order to achieve gasification reaction which are efficient and complete, the amount of the syngas in the outlet can be maximise by using feedstocks which has small particle size (diameter lower that 1 mm) (Serrano-Ruiz et al. 2010). Regarding the type of gasifier, large variety of designs have been patented in literature depending on feeding approaches (downdraft, updraft or direct entrained), bed configurations (fluidized or fixed), heat supply (indirect or direct) and working pressure (pressurised or atmospheric) (Milne et al. 1998). For applications of BTL, in has been indicated by research that they are not suitable to be operated at atmospheric pressure with direct air-blown type of gasifier because the stream of gas that were obtained is highly diluted in inert nitrogen. In contrast, the production of syngas is favour with the use of direct entrained type gasifier which allows processing at high temperatures (1500–1800 K), high pressures (10–60 bars) and also short residence times (Boerrigter and Van Der Drift 2004).

Cleaning of gas between the reactors is the main issue when the gasification of biomass was integrated with FTS. Apart from syngas, few contaminants were produced in the stream of gaseous from the gasifier, and these contaminants must be removed before it reaches the unit of Fischer-Tropsch which is very sensitive to impurities. In the technology of gasification, the main issue was represented by tars which is hydrocarbons that have high molecular weight, produced by incomplete gasification of biomass (Milne et al. 1998). Pipelines blockage and troublesome in operations will occur if the tars were condensed in the equipment of the downstream processing or in the gasifier. Selection of proper conditions of gasification and also design of the reactor will help in reducing the amount of tar produced (Devi et al. 2003), the gasification of the heavy hydrocarbons also can be assist by addition of solid catalysts based on Ru, Pt and Ni in the gasifier (Rapagna et al. 2000; Tomishige et al. 2004; Sutton et al. 2001; Mitchell and Brown 1985). Usually, the lignocellulosic biomass consists of variety of minor components for example inorganic materials based on potassium, halogens and also potassium, and proteins which is rich in sulphur. Therefore, the stream of gas that were produced from the gasification of biomass carries, NH_3, HCl, alkali and also volatile sulphur compounds which can cause corrosion on the turbines that was used for generation of electricity or, in the case of technology for BTL, these compounds can poison the catalysts that were used in the downstream unit of the Fischer-Tropsch. Furthermore, fine particles in the stream of gasification will clog the filters and also cause blockages, especially when the particle size of the feedstock used is small. For the reasons mentioned above, a unit of gas conditioning must be included between the reactor of the Fischer-Tropsch and the gasifier in the technologies of BTL. To develop an effective process of BTL, it has been noted that the key point is to have a sufficient gas cleaning (Ståhl et al. 2004).

Since the process of cleaning will involves some different types of contaminant such as chemicals, particles and also tars that have to be removed until ppm level (Spath and Dayton 2003), this unit usually consists of advanced technologies and multiple steps (Huber et al. 2006) which will significantly contribute to the cost and complexity of the plant of BTL. The other requirement for the composition of gas for units of Fischer-Tropsch is related to the molar ratio of CO/H_2. Usually, ratio of H_2/CO which is close to 2 are required for process of Fischer-Tropsch to hydrocarbon fuels taken place (Caldwell 1980; Dry 2002) and, because of the oxygen content in the biomass is high, common streams obtained from this resource have about 0.5 ratio of H_2/CO (Boerrigter et al. 2004b). Through reactions of water-gas shift (WGS), the ratio can be adjusted where the steam will react with the CO to produce H_2 and CO_2:

$$CO + H_2O \rightarrow CO_2 + H_2 \qquad (10.1)$$

In additional reactor of WGS, the adjustment can be carried out between the unit of Fischer-Tropsch and the gasifier, and also as alternative without the use of additional reactors, along with the biomass, extra water can be co-feed in the gasifier. However, the alternative method will give negative effects to the thermal efficiency of the gasification process. Lastly, the syngas that have been "shifted" and cleaned will be introduced into the reactor of FTS, which is the last unit of the BTL process. By using Fe-, Ru- or Co-based catalysts, FTS is a well-known process in industry that were used for production of alkanes (CnHn) from syngas (Dry 2002):

$$CO + 2H_2 \rightarrow (1/n)C_nH_n + H_2O \qquad (10.2)$$

The adjustment of the ratio of H_2/CO also can take place with reaction of WGS over catalysts of Fischer-Tropsch in the same synthesis bed. Poor selectivity in the final product of alkane is one of the main problems with the technologies of the Fischer-Tropsch, where the alkane distribution was broad in the range from C1 to C50.

The growth probability of alkane chain can be governed by the polymerization model of Anderson-Schulz-Flory (ASF), shows that neither diesel fuels nor gasoline can be selectively produced without producing undesired products in large amounts (Huber et al. 2006). To prevent this limitation, subsequent hydrocracking of these heavy compounds to diesel and gasoline and indirect routes which involves the initial production on heavy hydrocarbons (Dry 2004) and the use of active materials for isomerization and cracking, for example Fe and Co catalysts with ZSM-5 as their supports for production of gasoline components (Martínez and López 2005) are currently utilised for reactions of selective FTS. Currently, the activities of BTL are at the stages of development, research and also demonstration. In Netherlands, production of diesel fuel from woody biomass was demonstrated at one small plant (Boerrigter et al. 2004a), and at present time in Germany, a plant for production of 15,000 tons per year of liquid fuels from feedstocks of multiple lignocellulosic have been presented as the most promising projects of BTL.

Cost in production of fuel is one of the main issue with the commercialisation of technology of BTL, which was affected by the complexity of the process. Therefore, it is only economical to use this route on large scale, where the utilisation of large centralised facilities is required with the corresponding expense of the transportation of the biomass with low energy density. Along with the fuels of hydrocarbon, chemicals with high value such as methanol (Lange 2001) and hydrogen (Zhang et al. 2005) also were co-produced. The production of gas rich in H_2 by biomass gasification of steam can be obtained with in situ absorption of CO_2 in a system of dual fluidized bed. In order for improvement of economics for the process of BTL, biomass-derived syngas to fuel was used as the alternative.

10.3   Production of Renewable Hydrogen

Production of hydrogen in the form of molecular can be done in many different ways, from many different sources. In context of systems of energy, hydrogen is thought to be the best carrier of energy, and when compared to the fossil fuels that was extracted from the crust of earth, it is more akin to electricity. Hydrocarbon fuel also can be used to produce hydrogen, because these fuels also consists of hydrogen. Production of hydrogen also can be done by using water and other biological materials. When water is used for production of hydrogen, the process of "water-splitting" is called as electrolysis. And this process is known as the oldest process of electrochemical. Usually, the steam reformation of natural gas is utilised for production of hydrogen, besides that electrolysis also were used and hydrogen also produced as by-product of some processes of industry for example industry on production of chlor-alkali (Fig. 10.1).

10.3.1   Steam Methane Reforming

The process used for production of carbon dioxide and hydrogen over catalyst by reaction between steam and natural gas of other stream of methane such as landfill gas or biogas is called as steam methane reforming (SMR). For production of

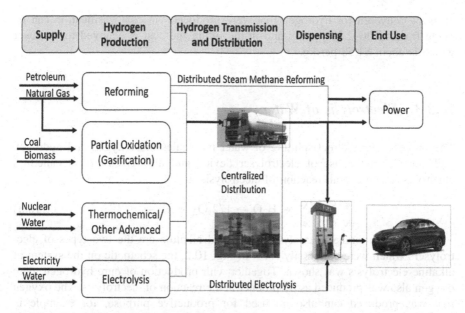

Fig. 10.1 Example of hydrogen production pathways

hydrogen on a basis of lower heating value, using natural gas as the starting material, the efficiency of SMR is approximately about 72%. When methane was used as the source, the efficiency of hydrogen production can be lower because methane consists of sulphur and other impurities that must be removed in the pre-treatment process before proceed to the upstream of the process of SMR. Production of gas which is rich in hydrogen by using SMR process is usually consist of about 70–75% of hydrogen, 6–14% of carbon dioxide, 7–10% of carbon monoxide and 2–6% of methane base on their dry mass (Hirschenhofer et al. 2000). The price of hydrogen produced by SMR might varies based on their cost of feedstock, production scale and many other variables. The cost of hydrogen can be range from $2–5 per kg at present. It is believed that the cost of delivered can be as low as $1.60 per kg in the future due to large centralized production and pipeline delivery.

10.3.2   Gasification of Coal and Other Hydrocarbons

Production of hydrogen using a range of hydrocarbon fuels, such as coal, low-value refinery products and also heavy residual oil can be carried out using process of partial oxidation which is also known as gasification. At temperature of 1200–1350 °C, a mixture of hydrogen and carbon monoxide can be produced by reaction between oxygen and hydrocarbon fuel in ratio that is less than stoichiometric. The costs of delivered for hydrogen that were produced via coal gasification can be range from $2.00–2.50 per kg, and in the future the cost is believed to be lower which is about $1.50 per kg.

10.3.3   Electrolysis of Water

The process where direct splitting of water molecules into oxygen and hydrogen molecules with the use of electrolyser device and also electricity is called as electrolysis. The overall reaction of electrolysis is:

$$e^- + H_2O \rightarrow 1/2\, O_2 + H_2 \qquad (10.3)$$

Polymer membrane electrolyte (PEM) and alkaline are the two types of electrolysers which were commonly used. In Fig. 10.2, the schematic on the system of alkaline electrolysis was shown. Together with production of pure hydrogen, pure oxygen also was produced as by-product in the reaction of electrolysis. The oxygen that was produced can also be used for productive purpose, for example in enriching the content of oxygen of greenhouses for production of food.

Production of hydrogen by water electrolysis can also be done using any source of electrical, such as wind power, nuclear power, hydropower, solar photovoltaic and also utility grid power. A wide range of scale, from few kW to 2,000 kW per

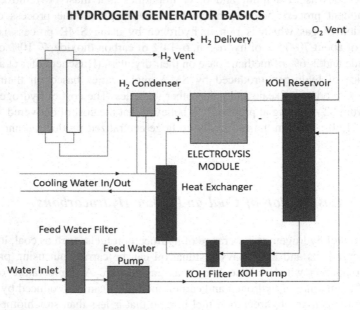

Fig. 10.2 Hydrogen production by alkaline electrolysis

electrolyser was used for electrolysis. It has been estimated that the delivered cost of hydrogen that was produced by grid power electrolysis will be around $6–7 per kg, and in the future, it is expected to be lower which is about $4 per/kg. For the cost of hydrogen that was produced by wind electrolysis, at present it will be about $7–11 per kg, and in the future, it will be lower which is $3–4 per kg. The hydrogen produced by solar could be more expensive, where presently the cost is $10–30 per kg and it is expected to be lowered to $3–4 per kg in the future.

10.3.4 Hydrogen from Biomass

Technologies for conversion of biomass can be divided into two, which are bio-chemical processes and also thermo-chemical processes. The thermos-chemical processes are cheaper because they can be operated at high temperature and thus the rate of reaction obtained will be higher. For production of gas-stream that is rich in hydrogen (syngas), they can involve either pyrolysis or gasification. A wide range types of biomass can be used. In contrast, presently, the biochemical processes which is enzyme-based are limited to sugar-based feedstocks and also wet. But in the future, the systems and the technique of process can still be improved. For medium scale production, the costs of delivered for hydrogen that was produced from biomass can be about $5–7 per kg. However, at larger scale production, the costs of delivered can be as low as $1.50–3.50 per kg. another option for production of hydrogen is by pyrolysis of biomass. The delivered costs of hydrogen produced by pyrolysis can also be low, as low as $1 per kg with production at large scale (Council 2004). However, to get those costs, production by pyrolysis at large scale are required, and it has not yet been realized on a commercial scale (Magrini-Bair et al. 2003). Besides that, to get hydrogen with high purity, the costs of clean up must be added in the production costs of hydrogen from biomass, and lastly costs of transports also must be included. These transport and purification costs can reach about additional of $1 per kg or more, depends on the scale of production and also any specific requirements.

10.3.5 High Temperature Fuel Cells

In order to directly run on methane, temperatures which are sufficiently high is required to operate the high temperature fuel cells based on solid oxide (SOFC) or molten carbonate (MCFC) technologies. It is also sometimes called as "internal reforming". Therefore, systems of SOFC and MCFC does not require stream of hydrogen which are relatively pure or pure as systems of PEM and PAFC. They can directly run on biogas, landfill gas or natural gas. Furthermore, the production of additional pure hydrogen as by-products can be designed in that systems, by feeding some additional fuel and also purifying the hydrogen-rich "anode tail gas" from the fuel cell to the purified hydrogen. Some clean-up of the stream of methane are required depends on the methane source. However, the success on using the systems of MCFC have been demonstrated by the projects in Chico, where the systems will run on a blend of natural gas and also brewery treatment of wastewater digester gas. Currently in California, the Public Utilities Commission Self-Generation Incentive Program (SGIP) provides an installation incentive of $2.50 per Watt for systems of fuel cell running on natural gas, and $4.50 per Watt for systems of fuel cells running on a renewable fuel.

10.3.6 Other Methods of Hydrogen Production

Production of hydrogen also can be done by other ways, such as direct solar process of electrochemical, algae and also various pathways with nuclear-power-assisted (Lipman 2004). However, in recent years, some additional progress in these areas has been done significantly, and some of them tend to be promising to be use in production of hydrogen using renewable sources, together with lower costs, less loss of energy and also less emissions of greenhouse gases when compared to currently typical of industry practice.

References

Agrawal R, Singh NR, Ribeiro FH, Delgass WN (2007) Sustainable fuel for the transportation sector. Proc Natl Acad Sci 104(12):4828–4833

Baliban RC, Elia JA, Floudas CA (2010) Toward novel hybrid biomass, coal, and natural gas processes for satisfying current transportation fuel demands, 1: process alternatives, gasification modeling, process simulation, and economic analysis. Ind Eng Chem Res 49 (16):7343–7370

Baliban RC, Elia JA, Floudas CA (2011) Optimization framework for the simultaneous process synthesis, heat and power integration of a thermochemical hybrid biomass, coal, and natural gas facility. Comput Chem Eng 35(9):1647–1690

Baliban RC, Elia JA, Floudas CA (2012) Simultaneous process synthesis, heat, power, and water integration of thermochemical hybrid biomass, coal, and natural gas facilities. Comput Chem Eng 37:297–327

Boerrigter H, Van Der Drift A (2004) Biosyngas: description of R&D trajectory necessary to reach large-scale implementation of renewable syngas from biomass. Energy Research Centre of the Netherlands, The Netherlands

Boerrigter H, Calis HP, Slort DJ, Bodenstaff H (2004a) Gas cleaning for integrated biomass gasification (BG) and Fischer-Tropsch (FT) systems; experimental demonstration of two BG-FT systems. Acknowledgement/Preface:51

Boerrigter H, Zwart R, van Ree R, Veringa H (2004b) High efficiency co-production of Fischer-Tropsch (FT) transportation fuels and Substitute Natural Gas (SNG) from biomass. ECN, The Netherlands

Borgwardt RH (1997) Biomass and natural gas as co-feedstocks for production of fuel for fuel-cell vehicles. Biomass Bioenerg 12(5):333–345

Bridgwater AV, Double JM (1991) Production costs of liquid fuels from biomass. Fuel 70 (10):1209–1224

Bridgwater A, Czernik S, Piskorz J (2001) An overview of fast pyrolysis. In: Progress in thermochemical biomass conversion, pp 977–997

Caldwell L (1980) Selectivity in Fischer-Tropsch synthesis: review and recommendations for further work

Chen Y, Adams TA, Barton PI (2011) Optimal design and operation of flexible energy polygeneration systems. Ind Eng Chem Res 50(8):4553–4566

Clausen LR, Elmegaard B, Houbak N (2010) Technoeconomic analysis of a low CO_2 emission dimethyl ether (DME) plant based on gasification of torrefied biomass. Energy 35(12): 4831–4842

Council NR (2004) The hydrogen economy: opportunities, costs, barriers, and R&D needs. National Academies Press, Washington D.C.

Council NR (2008) Water implications of biofuels production in the United States. National Academies Press, Washington D.C.

Council NR (2010) Liquid transportation fuels from coal and biomass: technological status, costs, and environmental impacts. National Academies Press, Washington D.C.

Devi L, Ptasinski KJ, Janssen FJ (2003) A review of the primary measures for tar elimination in biomass gasification processes. Biomass Bioenerg 24(2):125–140

Dong Y, Steinberg M (1997) Hynol—an economical process for methanol production from biomass and natural gas with reduced CO_2 emission. Int J Hydrogen Energy 22(10–11): 971–977

Dry ME (2002) The fischer–tropsch process: 1950–2000. Catal Today 71(3):227–241

Dry M (2004) FT catalysts. Stud Surf Sci Catal 152:533–600

Elia JA, Baliban RC, Xiao X, Floudas CA (2011) Optimal energy supply network determination and life cycle analysis for hybrid coal, biomass, and natural gas to liquid (CBGTL) plants using carbon-based hydrogen production. Comput Chem Eng 35(8):1399–1430

Elia JA, Baliban RC, Floudas CA (2012) Nationwide energy supply chain analysis for hybrid feedstock processes with significant CO_2 emissions reduction. AIChE J 58(7):2142–2154

Floudas CA, Elia JA, Baliban RC (2012) Hybrid and single feedstock energy processes for liquid transportation fuels: a critical review. Comput Chem Eng 41:24–51

Hamelinck CN, Faaij AP, den Uil H, Boerrigter H (2004) Production of FT transportation fuels from biomass; technical options, process analysis and optimisation, and development potential. Energy 29(11):1743–1771

Henrich E, Dahmen N, Dinjus E (2009) Cost estimate for biosynfuel production via biosyncrude gasification. Biofuels, Bioprod Biorefin 3(1):28–41

Hirschenhofer J, Stauffer D, Engleman R, Klett M (2000) Fuel cell handbook. Business/Technology Books

Huber GW, Iborra S, Corma A (2006) Synthesis of transportation fuels from biomass: chemistry, catalysts, and engineering. Chem Rev 106(9):4044–4098

Kreutz TG, Larson ED, Liu G (2008) Williams RH Fischer-Tropsch fuels from coal and biomass. In: 25th annual international Pittsburgh coal conference. vol 2.10, Citeseer

Lange J-P (2001) Methanol synthesis: a short review of technology improvements. Catal Today 64 (1):3–8

Lange JP (2007) Lignocellulose conversion: an introduction to chemistry, process and economics. Biofuels, Bioprod Biorefin 1(1):39–48

Larson ED, Jin H, Celik FE (2009) Large-scale gasification-based coproduction of fuels and electricity from switchgrass. Biofuels, Bioprod Biorefin 3(2):174–194

Larson ED, Fiorese G, Liu G, Williams RH, Kreutz TG, Consonni S (2010) Co-production of decarbonized synfuels and electricity from coal+ biomass with CO_2 capture and storage: an Illinois case study. Energy Environ Sci 3(1):28–42

Laser M, Jin H, Jayawardhana K, Dale BE, Lynd LR (2009a) Projected mature technology scenarios for conversion of cellulosic biomass to ethanol with coproduction thermochemical fuels, power, and/or animal feed protein. Biofuels, Bioprod Biorefin 3(2):231–246

Laser M, Larson E, Dale B, Wang M, Greene N, Lynd LR (2009b) Comparative analysis of efficiency, environmental impact, and process economics for mature biomass refining scenarios. Biofuels, Bioprod Biorefin 3(2):247–270

Li H, Hong H, Jin H, Cai R (2010) Analysis of a feasible polygeneration system for power and methanol production taking natural gas and biomass as materials. Appl Energy 87(9):2846–2853

Lipman TE (2004) What will power the hydrogen economy? Present and future sources of hydrogen energy

Liu G, Larson ED, Williams RH, Kreutz TG, Guo X (2010) Making Fischer – Tropsch fuels and electricity from coal and biomass: performance and cost analysis. Energy Fuels 25(1):415–437

Liu G, Williams RH, Larson ED, Kreutz TG (2011a) Design/economics of low-carbon power generation from natural gas and biomass with synthetic fuels co-production. Energy Procedia 4:1989–1996

Liu P, Whitaker A, Pistikopoulos EN, Li Z (2011b) A mixed-integer programming approach to strategic planning of chemical centres: a case study in the UK. Comput Chem Eng 35(8):1359–1373

Magrini-Bair KA, Czernik S, French R, Chornet E (2003) Fluidizable catalysts for hydrogen production from biomass pyrolysis/steam reforming. FY 2003 Progress Report, National Renewable Energy Laboratory

Martínez A, López C (2005) The influence of ZSM-5 zeolite composition and crystal size on the in situ conversion of Fischer-Tropsch products over hybrid catalysts. Appl Catal A 294 (2):251–259

Milne TA, Abatzoglou N, Evans RJ (1998) Biomass gasifier" tars": their nature, formation, and conversion, vol 570. National Renewable Energy Laboratory Golden, CO

Mitchell D, Brown M (1985) Catalytic steam gasification of biomass for methanol and methane production1. J Sol Energy Eng 107:89

Nashawi IS, Malallah A, Al-Bisharah M (2010) Forecasting world crude oil production using multicyclic Hubbert model. Energy Fuels 24(3):1788

Onel O, Niziolek AM, Floudas CA (2016) Optimal production of light olefins from Natural gas via the methanol intermediate. Ind Eng Chem Res 55(11):3043–3063

Perales AV, Valle CR, Ollero P, Gómez-Barea A (2011) Technoeconomic assessment of ethanol production via thermochemical conversion of biomass by entrained flow gasification. Energy 36(7):4097–4108

Perlack R, Wright L, Turhollow A, Graham R, Stokes B, Erbach D (2005) A bioenergy and bioproducts industry: the technical feasibility of a billion ton annual supply. US Department of Energy, Office of Scientific and Technical Information, Po Box 63:37831–30062

Rapagna S, Jand N, Kiennemann A, Foscolo P (2000) Steam-gasification of biomass in a fluidised-bed of olivine particles. Biomass Bioenerg 19(3):187–197

Serrano-Ruiz JC, Braden DJ, West RM, Dumesic JA (2010) Conversion of cellulose to hydrocarbon fuels by progressive removal of oxygen. Appl Catal B 100(1):184–189

Sharma P, Sarker B, Romagnoli JA (2011) A decision support tool for strategic planning of sustainable biorefineries. Comput Chem Eng 35(9):1767–1781

Spath PL, Dayton DC (2003) Preliminary screening-technical and economic assessment of synthesis gas to fuels and chemicals with emphasis on the potential for biomass-derived syngas. DTIC Document

Ståhl K, Waldheim L, Morris M, Johnsson U (2004) Gårdmark L Biomass IGCC at Värnamo. GCEP energy workshop, Sweden-past and future. In

Sunde K, Brekke A, Solberg B (2011) Environmental impacts and costs of hydrotreated vegetable oils, transesterified lipids and woody BTL—a review. Energies 4(6):845–877

Sutton D, Kelleher B, Ross JR (2001) Review of literature on catalysts for biomass gasification. Fuel Process Technol 73(3):155–173

Tijmensen MJ, Faaij AP, Hamelinck CN, van Hardeveld MR (2002) Exploration of the possibilities for production of Fischer Tropsch liquids and power via biomass gasification. Biomass Bioenerg 23(2):129–152

Tock L, Gassner M, Maréchal F (2010) Thermochemical production of liquid fuels from biomass: thermo-economic modeling, process design and process integration analysis. Biomass Bioenerg 34(12):1838–1854

Tomishige K, Asadullah M, Kunimori K (2004) Syngas production by biomass gasification using Rh/CeO$_2$/SiO$_2$ catalysts and fluidized bed reactor. Catal Today 89(4):389–403

Wang J, Feng L, Tverberg GE (2013) An analysis of China's coal supply and its impact on China's future economic growth. Energy Policy 57:542–551

Warren A, El-Halwagi M (1996) An economic study for the co-generation of liquid fuel and hydrogen from coal and municipal solid waste. Fuel Process Technol 49(1–3):157–166

Williams RH, Liu G, Kreutz TG, Larson ED (2011) Alternatives for decarbonizing existing USA coal power plant sites. Energy Procedia 4:1843–1850

Wright MM, Daugaard DE, Satrio JA, Brown RC (2010) Techno-economic analysis of biomass fast pyrolysis to transportation fuels. Fuel 89:S2–S10

Zhang R, Cummer K, Suby A, Brown RC (2005) Biomass-derived hydrogen from an air-blown gasifier. Fuel Process Technol 86(8):861–874

Chapter 11
Catalytic Transformation of CO$_2$ to Fuels

11.1 Introduction

By the year 2050, the European Commission are committed to reduce the emissions of greenhouse gas to 80–95% through the "Energy Roadmap 2050" that was adopted on December 15, 2011 (Schleicher-Tappeser 2012). The routes on system of energy for decarbonization was explored by this Energy Roadmap 2050, and in order to achieve this objective, many relevant contributions have been done, which has effects not only at the European but also worldwide level. For these scenarios, biofuels are one of the element which is relevant as part of the objective in realizing a sustainable bio-economy. Reducing the emissions of CO$_2$ also categorise as sustainable, including the possibility of using CO$_2$ (Quadrelli et al. 2011; Centi and Perathoner 2009; Centi et al. 2013). Many different motivations have been pushing towards the utilisation of biofuels, lead to the fact that for production of liquid biofuels, this is the only sustainable way, which are and will stay to be the vectors of energy that is important for transportation industry which accounting about one-third of the global energy consumption. Due to the number of vehicles circulating in emerging countries are increasing, the fast expansion in the utilisation of liquid fuels will not be compensated by the introduction of devices which is more-efficient. Thus, for the next 2 or 3 decades, transport based on carbon footprint can be reduced by introduction of fuels which are derived from renewable resources. However, when compared to fuels which were derived from fossil resources, the biofuels production is more expensive. Besides that, for production to second and third-generation from first generation of biofuels, progressive passage is needed. The production cost also will probably be higher in order to reduce the impact on social and environment, and also to avoid competition with food.

For production of biofuels, usually it will depend largely on subsidies which are motivated by different reasons, such as the objective to reduce the impact of greenhouse gas of the transport sector, while the actual energy is preserved. Therefore, there is an obvious relationship between the emission of CO$_2$ and the

© Springer International Publishing AG 2017   191
S. Bagheri, *Catalysis for Green Energy and Technology*,
Green Energy and Technology, DOI 10.1007/978-3-319-43104-8_11

biofuels. In production of biofuel, the effective effects on emissions of greenhouse gas have been discussed, and must be considered. As indicated by life-cycle analysis (LCA), effective contribution of biofuels can be contributed by the reduction in emissions of CO_2 that are negative and close to neutral. Presently, some factors in schemes of sustainability have been investigated, including emissions of soil N_2O and change in indirect land use, and it is not easy to predict the effect of these factor (Cherubini and Jungmeier 2010). The type of biofuels and their raw materials will affects the result considerably (Zinoviev et al. 2010). When compared to the use of fuels derived from fossil, reduction in carbon footprint of mobility can be achieved by biofuels. However, any possibility in reducing the carbon footprint in biofuels production and introducing another renewable source in this chain of energy must be considered.

In fact, current methods on production of renewable energy (solar, geo, wind and hydro) produce electrical energy essentially, in which their main limits are the need of a distribution grid and storage with low efficiency. In biorefinery, the emissions of greenhouse gas were minorly effected by direct usage of renewable electrical energy, because from total effect, only few percentages were associated with the use of electrical energy that was derived from fossil fuels. Besides that, CO_2 production is another aspect that was rarely considered in the biofuels production. For example, let us discuss the production of bioethanol. Production of ethanol can be carried out by fermentation of sugar crops and various grains such as barley, corn, wheat and sorghum, with CO_2 produced as a by-product. The main difference in the process of second-generation bioethanol derived from sources of lignocellulosic is just the pre-treatment of the biomass, where the process of fermentation remains the same. Production of bioethanol can be done by fermentation of six-carbon sugar such as glucose by yeast. Conversion of glucose into CO_2 and ethanol occurs during the fermentation as shown in Eq. 11.1:

$$C_6H_{12}O_6 \rightarrow 2C_2H_5OH + 2CO_2 \qquad (11.1)$$

Approximately, almost 1 t CO_2 was emitted per ton of ethanol, due to their similar molecular weight. According to IEA regarding the rise scenario of 2 °C (Lantz et al. 2012), in 2020, the utilisation of biofuel is expected to increase to about 240 billion litres, in order to achieve about 0.1 Gt reduction of emissions of CO_2 in the sector of transport. However, the process of fermentation as shown in Eq. (11.1) would results in production of CO_2 at about 0.23 Gt. For fuels or chemicals production using renewable energy, bioethanol can significantly improve the effective positive impact on emissions of greenhouse gas. The effects are double positive: avoid the CO_2 emissions and renewable energy will be introduce directly in the sector of transport by its utilisation in conversion of liquid fuels from CO_2 (Centi and Perathoner 2010, 2011; Centi et al. 2011b). Therefore, in this case, double-win scenario can be observed, where in order to foster the transition to economy with low-carbon footprint, effective strategy by recycle CO_2 to fuels were done and in a longer term to a solar fuel economy.

11.2   CO$_2$ Emissions in Biorefinery

In particular, biorefinery is a platform to prepare molecules for integrated approach in production of fuels or chemicals, which use biomass of lignocellulosic as raw material (Cherubini and Strømman 2011; Centi et al. 2011a; Mandl 2010; Wellisch et al. 2010). The typical case for refinery is the production of bioethanol. The issues of adding value to CO$_2$ from fermentations of ethanol have been reviewed by Xu et al. (2010), and due to the exponential increase in the production of ethanol, the relevance of this subject was increasing.

In 2008, the emissions of CO$_2$ associated with fermentation of CO$_2$ for production of bioethanol were about 50 MMT (Xu et al. 2010) and in 2011 it has reached about 70 MMT. A simplified process of fermentation of bioethanol is shown in Fig. 11.1.

For various steps, both different technologies and also different configuration are possible (Xu et al. 2010; Piccolo and Bezzo 2009). However, this is not relevant to

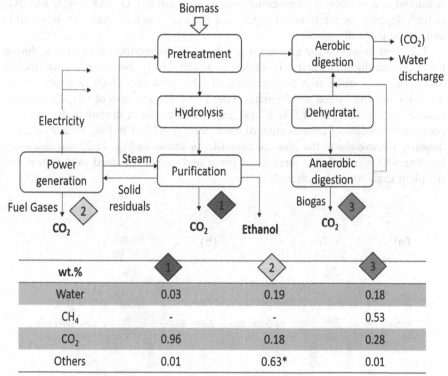

wt.%	1	2	3
Water	0.03	0.19	0.18
CH$_4$	-	-	0.53
CO$_2$	0.96	0.18	0.28
Others	0.01	0.63*	0.01

*N$_2$ and residuals O$_2$ (from combustion air), CO, pollutants (NOx, SOx, VOC), ashes

Fig. 11.1 Block flowsheet of a general biomass to ethanol process in a biorefinery, with integrated power generation and aerobic/anaerobic treatment units for wastewater. The table below the block flowsheet reports the composition of CO$_2$, methane, and water in some the streams indicated with the same code in the block flowsheet

illustrate the opportunities which are related to the utilisation of CO_2. The flowsheet of basic process are illustrated in Fig. 11.1, which includes the utilisation of solid residuals from biomass for steam generation that were utilised in the separation and also pre-treatment units, and in part to cogenerate the electrical energy utilised in utilities of process (Piccolo and Bezzo 2009). For production of biogas, stream of aqueous which consist of significant amounts of organics and also suspended solids are sent to a first unit of anaerobic digestion, while the residual of organic matter will be send to first unit of aerobic digestion, and finally they will be discharged. The relative composition (wt%) and also the main sources of CO_2 from the process was shown in Fig. 11.1. Typically, the CO_2 emissions from the aerobic digestion was not collected, therefore they are not included. In the process of bioethanol, there are three main points of emissions of CO_2. The first points are related to the step of fermentation, where the CO_2 produced will have high-purity which is more than 95% and is also more relevant quantitatively. The second points are related to the solid residuals combustion where the steam which is required in the process will be generated. The CO_2 concentration in the gas of flue is typically below 20%, and is related to a number of other components, from ash to CO, VOC, NO_x and SO_x (which depends on the biomass type), and it makes the CO_2 recovery is possible and uneconomically used.

The third points on the emissions is related to the production of biogas during the anaerobic digestion, that is less relevant quantitatively. However, it is of interest due to its association with the amounts of CH_4 presence which depends on the digestion conditions and also biomass type. Typically, the ratio of $CH_4:CO_2$ in any case is in the range of 1.5–0.8. the mass balance for a common dry-mill for production process of corn to ethanol have been simplified in Fig. 11.2 in order to quantify the streams in the plant of ethanol. As shown in Fig. 11.2, two cases can be observed with different streams of input and output: (a) feed of 1 t corn and (b) plant capacity of 1 t ethanol.

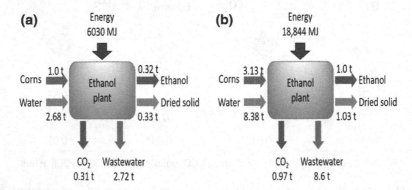

Fig. 11.2 Simplified mass balance for a typical dry-mill corn-to-ethanol production process, to quantify the different input-output streams for two cases Dewil et al. (2006). **a** 1 t corn feed and **b** 1 t ethanol plant capacity

11.3   Current Uses of CO_2 Emissions in Ethanol Production by Fermentation Plants

For many years, the captured CO_2 from the processes of fermentation has contributed a quite small but crucial part in the market of CO_2 merchant (Supekar and Skerlos 2014). For instance, distilleries, breweries and fuel ethanol plants are commonly used to capture CO_2 gas which was generated as a by-product in operations for production of liquid CO_2 with high-purity. Annually, the market of CO_2 merchant is around 20 MMT, where the synthesis of urea was not included, because in the latter case of CO_2 which was derived directly from the H_2 that was synthesised from methane. Ammonia, which is the other reactant of urea can be produced by utilisation of H_2. In the industry of food and associate such as processing of vegetables, freezing of specialty food, bottling of soft drink, and packaging of modified atmosphere applications, mainly CO_2 were used. Currently, for this market, because of the CO_2 emitted from the plants of ethanol have high-purity, about one-third was derived from the processes of production of ethanol by fermentation.

Other CO_2 sources for the market were derived from reformer of hydrogen, production of anhydrous ammonia, natural wells, titanium oxide and ethylene oxide plants. Annually, when compared to the emissions of CO_2 which is related to fermentation in plants of bioethanol, the growth of this market is much lower even though it is in expansion for about 1–3%. The CO_2 also can be used directly to stimulate the gas wells, EOR and also enhance the removal of CBM. However, these uses are sensitive to the fossil fuels value, commonly pipeline delivered and also specific to markets. Unlike the business merchant end, where the transportation of CO_2 as ice is using either railcars or trailers. Therefore, in plants of bioethanol, it has been forecasted that the amount of CO_2 that is not used was increased progressively (Supekar and Skerlos 2014).

11.4   Composition of CO_2 Emissions in Ethanol Plants

The production of high-purity CO_2 ($\sim 99\%$) can be obtained by fermentation. In the stream of CO_2, the main contaminants are water, air, alcohols (especially ethanol), ketones, aldehydes and also sulphur compounds. The types of process and feedstock will affects the amount of contaminants (Woods et al. 2008). A series steps of purification is required to remove the contaminants in order to be utilised as food. Particularly, to be used as fuels, no severe problems will be created by these contaminants. By using the conventional technologies of sulphur-removal, the low amount of sulphur compounds can be removed, for instance, the natural gas must be purified before the steam reforming or the H_2 must be purified for fuel cells (De Wild et al. 2006). The performances of catalyst and also the quality of the fuel products will not be affected by the presence of traces contaminants. Usually, scrubbing can be used to remove these contaminants traces easily.

To be utilised for food and other related industry, it is not always economical when the sources of fermentation is used to produced CO_2, and many breweries, for instance, for production of beer from external sources the purchase of CO_2 is required, rather than the generated gas was captured in their own fermenters, the emitted CO_2 from the plants of production of bioethanol is an opportunity which is potentially valuable for used of chemical. Composition of biogas which is derived from the anaerobic digestion as shown in Fig. 11.1, depends on the specific operative and biomass conditions, as well as the microorganism type. Generally, it consists of CO_2, methane, siloxanes trace, ammonia, hydrocarbons, water and also hydrogen sulphide (Ajhar et al. 2010; McBean 2008; Dewil et al. 2006). In batch systems, the composition also will vary by time. However, in the continuous system, the compositions are fluctuated and it depends on the changes in pH, temperatures and also population of microorganisms (Osorio and Torres 2009).

Methane is the main product along with carbon dioxide. When compared to former case which is CO_2 from fermentation, this is their main difference, because the hydrogen sources that can be used for conversion of CO_2 is already there, for instance, by wet reforming. On the other hand, it must be asked whether, on basis of energy, is it worthy to convert CO_2 from methane or is it more preferred to simply use methane as SNG and eliminate the CO_2. With respect to fermentation of CO_2, the impurities with complex composition is another difference between biogas, consists of elements such as siloxanes in low amount (Ajhar et al. 2010; McBean 2008; Dewil et al. 2006), the catalysts deactivation can occur by fouling. Siloxanes is a family of man-made organic compounds that consists of oxygen, silicon and also methyl groups. It can be found in the waste utilised for generation of biogas and the siloxanes with low molecular weight will be volatilise into digester gas. Transformation of siloxanes to silicate oxides will occurs at high temperatures and it can cause the damage of the equipment and catalyst. The presence of dust also can occur. And by using filters based such as cloth of activated carbon, removal of both can be carried out (Ajhar et al. 2010). Before the CO_2 is reacted with methane, purification of the raw biogas that was produced in the digester must be done, by removing the ammonia, sulphur compounds, siloxanes and also the other organic impurities. The technologies for purification are available (Ryckebosch et al. 2011), however they are costly since the removal of contaminants must be carried out without eliminating the CO_2.

11.5   Options for Using CO_2 Emissions in Ethanol Plants

As discussed in the latter case, due to its purity which requires minimum treatment to be utilised, the CO_2 from fermentation is a valuable source. However, it is less clear for the second case which involves the presence of CO_2 in biogas. For utilisation of biogas, there are several different ways which are possible, for example steam and heat production on-site, generation or cogeneration of electricity, use for vehicle fuel, injection in the grid as SNG, and also chemicals production. However, because of the purification process is costly, it is indicated as an option which are less valuable (Appels et al. 2008). Typically, elimination of CO_2 from the stream is required for the other purposes, and it can be done by either adsorption or absorption. In fact, the characteristics of flame will be affected negatively when the CO_2 is present at significant amount in the mixture with methane. The lower efficiency of combustion can occur when the rates of burning and also the temperatures of flame were reduced, and the stability of flame is in a narrower range. Furthermore, production of NO_x per gram of methane can be increased with high CO_2 concentration. After purification, the biogas can be utilised directly for production of syngas and then energy or chemicals vectors by using the syngas which was derived from biogas. Production of syngas was carried out by reaction of methane with CO_2 in the presence of catalyst via dry reforming:

$$CH_4 + CO_2 \rightarrow 2CO + 2H_2 \quad \Delta H298 = +247\,kJ/mol \qquad (11.2)$$

The reaction of dry reforming is highly endothermic, and therefore the excess part of CH_4 in the biogas must be oxidised as shown in Eq. 11.3, for production of heat required in reaction to sustain the methane wet reforming as shown in Eq. 11.4 which is also endothermic and also dry reforming. The reaction of WGS also present (Eq. 11.5), as well as the combustion of methane (Eq. 11.6).

$$CH_4 + 0{:}5O_2 \rightarrow CO + 2H_2 \quad \Delta H298 = -36\,kJ/mol \qquad (11.3)$$

$$CH_4 + H_2O \rightarrow CO + 3H_2 \quad \Delta H298 = +206\,kJ/mol \qquad (11.4)$$

$$CO + H_2O \rightarrow CO_2 + H_2 \quad \Delta H298 = -41\,kJ/mol \qquad (11.5)$$

$$CH_4 + 2O_2 \rightarrow CO_2 + 2H_2O \quad \Delta H298 = -880\,kJ/mol \qquad (11.6)$$

Tri-reforming is referred to the combination of these reactions. Typically, in biogas where methane is in excess with respect to CO_2, a syngas with ratios of H_2: CO is less than 1, suitable for methanol synthesis and production of DME. Even though various problems such as poisoning of sulphur, activity, formation of carbon and sintering are present (Mota et al. 2011), the tri-reforming of biogas that was produced from biomass anaerobic digestion is technically a feasible solution for syngas production, which can be further converted to energy vectors and also chemicals. Production of hydrogen by reforming of biogas have been explored by many authors (Xu et al. 2009) or as feed to fuel cells of solid oxide. It is very useful for the reactions of thermodynamic and the enthalpies of relevant reactions to be summarised in order to have a picture of the problem that is more complete as shown in Fig. 11.3 (Jiang et al. 2010). There are two different options that needs to be clarified: biogas conversion to energy vectors or chemicals for the energy of transport or storage, and the use of biogas directly for applications of energy.

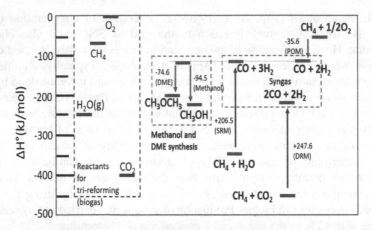

Fig. 11.3 The enthalpy of reaction for syngas production and Fischer-Tropsch (*FT*) synthesis of methanol and dimethyl ether. *DRM* dry reforming of methane with CO_2; *SRM* steam reforming of methane; *POM* partial oxidation of methane. Adapted from Mota et al. (2011)

Let's only consider the second case, because the only option which is reasonable in the first case is the use of biogas directly after purification and CO_2 removal. The same is true for the utilisation of biogas as SNG. For the second case (conversion to energy vectors or chemicals), the syngas must be converted from biogas, which can be converted to methanol, FT products, olefins or DME. These different products type not only differ in relation to utilise, but also in terms of the ratio of H_2/CO required in the synthesis of syngas. Particularly, the stoichiometric ratio for DME or methanol is 2.0, while for olefin and FT syntheses using a stream of pure CO/H_2 the ratio is lower. However, some of the CO_2 can be co-fed and the reaction of WGS as shown in Eq. 11.5 is always present in all of these reactions. Therefore, there will be slight different in the optimal ratio which depends on the composition of feed, catalyst and also the conditions of reaction. In the DRM, its stoichiometric ration of $H_2:CO$ is 1.0, while 3 and 2 in the SRM and POM, respectively.

In the reaction of DRM, the conversion of CO to CO_2 is required and can be carried out by WGS, and then the CO_2 have to be removed to reach the required ratio of $H_2:CO$ for the consecutive syntheses. The DRM reaction is highly endothermic, and when compared to the SRM reaction, DRM is even more endothermic, and therefore the methane which is excess in the biogas must be utilised to sustain the reaction by POM, but especially by the combustion reactions. Therefore, even though the reaction of DRM allows a better utilisation of all carbon which is present in the biogas, with respect to the alternative CO_2 removing from biogas, the efficient utilisation is not better or even worse, and then utilisation of methane foe partial oxidation or steam reforming reactions. In recent years, a large interest of research on the of reforming of biogas has been presented in literature (Lau et al. 2012; Chang et al. 2012; da Silva et al. 2012), with biogas is considered as valuable source of renewable energy and for production of syngas or H_2 (Mota et al. 2011). However, when compared to the use of only methane, no advantage on the combined utilisation of CO$_2$ and methane present in the biogas have been really demonstrated by these studies.

A study on different process for production of methanol based on sources of renewable energy have been analysed by Clausen et al. (2010a) recently. For catalytic synthesis of methanol, the syngas utilised was produced by different routes. The three relevance schemes of process are: methane autothermal reforming with separate CO$_2$ addition, biogas reforming, and direct use of carbon dioxide. In all the three cases, water electrolysis is utilised to produce hydrogen (Clausen et al. 2010b). It can be concluded that, in biorefineries, the CO$_2$ from fermentation is a possible source of carbon which is valuable for valorisation. While, other emissions point is actually of less interest, except when high taxes on emissions of carbon will be applied.

Specifically, how the integration of microalgae in the biorefineries of ethanol to reduce the emissions of carbon dioxide have been discussed by Rosenberg et al. (2011). In the discussion, a case in Iowa on biorefineries of 50-MGY ethanol have been illustrated. In terms of land-use and cost, negative conclusion was drawn. However, how the integration of the two approaches presents the issues number and cannot be realised straight have been pointed out. Approaches which are more complex have been explored for this reason. For instance, the couple of yeast fermenters of a plant of bioethanol with microalgae column photobioreactors to generate fuel cells of coupled microbial have been proposed by Powell and Hill (2009).

11.6   CO$_2$ to Valorise Waste and Produce High-Value-Added Chemicals

For production of methanol, the utilisation of CO$_2$ in biorefinery, for instance, the possibility to explore the utilisation of methanol for by-products of refinery functionalization have been suggested in order to synthesise the chemicals which is high-value-added. The advantage of using carbon dioxide can be joined by this strategy with the possibility by utilising a by-product of biorefinery to synthesise a fine chemical which is valuable in where the economics of entire process can be improved. The following example illustrates that this is a possibility, therefore joining production of biorefinery and creative CO$_2$ utilisation in a cycle of virtuous.

Furfural is a by-product in production of second-generation bioethanol from sources of lignocellulosic (Demirbas 2008). It can be produced through dehydrating five-carbon sugars for example arabinose and xylose that were derived from the biomass fraction hemicellulose. Furfural must be removed due to its effects of inhibitory on growth of microbial and metabolism in the fermentation of bioethanol (Boyer et al. 1992). In biorefinery, furfural will finds the application, such as, to produce valerate esters, ethylfurfuryl, 2-methylfuran, methyltetrahydrofuran, ethyltetrahydrofurfuryl ethers and also various products of C10–C15 coupling

(Lange et al. 2012). Furfural is an oily, almond-scented and colourless liquid that will turns into dark brown from yellow when they are exposed to air. Furfural also can be used as a fungicide, as a solvent in lube oil refining, weed killer, tetrahydrofuran production and important solvent in industry. However, all these are chemicals with a relatively low-value and it is therefore interesting to upgrade these chemicals into chemicals with higher-value. The potential market is huge with the value of global market is almost $20 billion for utilisation of furfural as perfumes, agents of food-flavouring and also fragrances. The fragrances of furan-based are available in wide range in many ingredients which is natural. As shown in Fig. 11.4, production of methyl-2-furoate is possible by selective oxidative methylation reaction between furfural and methanol.

Furfural is ingredients contained in flavours and fragrances that was utilised in decorative cosmetics, shampoo, toilet soaps, fine fragrances and other toiletries, in flavours of oral care products and also in the products of non-cosmetic, for example, detergents and cleaners for household. However, it is labelled as H351 indicated as the suspect of causing cancer.

Even though conclusion have been made by specific investigation that there is no significant risk of cancer at the doses used, recently, the normative indicated that it is preferable if the furfural was substituted with other compounds. The risks which is related to the utilisation of furfural directly can be reduced by its conversion to less risky and alternative agents of flavouring and fragrances. At room temperature, in methanol, oxidation of furfural to methyl-2-furoate using 1 bar oxygen on gold catalyst was carried out and have been observed by Taarning et al. (2008). Since there is only aldehyde moiety presence as the only functional group, under reaction conditions that is very mild (20 °C) the full oxidation to methyl furoate can be done in a relatively short time. Distillation process can be carried out to purify the formed methyl furoate. In the presence of strong base, which has to be neutralised in the stream of product after the reaction, the results were obtained by Taarning et al. (2008). The utilisation of strong base must be avoided in order to have a better process which is sustainable. Recently, high selectivity and activity

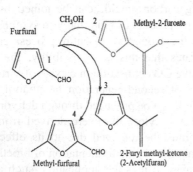

2. Methyl-2-furoate (Nutty, peppermint, tobacco odor with mushroom undertones; Use: Perfuming agent for an earthy, tobacco, minty note in specialty fragrances)

3. 2-Acetylfuran (Odor: balsamic, caramel, sweet; Use: flavouring in bakeries, chocolate, cocoa, coffee, nut, tomato)

4. Methyl-furfural (Odor: nutty, caramel, roast burnt; Use: fragrance agent in alcoholic beverages, bread, caramel, cranberry, tropical fruits

Fig. 11.4 Synthesis of fragrances and flavouring agents by catalytic reaction of furfural with methanol and characteristics of the products obtained

has been shown by Au/ZrO$_2$ catalyst in the conversion of methyl furoate from furfural by oxidative esterification with methanol and oxygen, without any base has been showed by Pinna et al. (Menegazzo et al. 2008). In order to get high selectivity which is more than 90%, the particle size of gold (<2 nm) is the critical parameter that need to be controlled.

11.7   Catalytic Conversion of CO$_2$

Different possibilities for utilisation of CO$_2$ in biorefinery have been discussed previously and currently, it has been evidenced that the production of H$_2$ by CO$_2$ catalytic conversion from renewable sources is the only possible solution, even though there will be more integrated solution will be possible in the future. Therefore, it is worthy to overview the art state in the CO$_2$ catalytic conversion using H$_2$. The discussion was limited only to methanol and its conversion product (DME).

11.7.1   New Aspects in the Synthesis of Methanol from CO$_2$

Commercially, the production of methanol was carried out from syngas. The reaction will be accelerated with the presence of CO$_2$ in the feed. Recently, fixation of CO$_2$ into methanol at the interface of Cu/ZrO$_2$ have been analysed from Monte Carlo and first principle approaches (Tang et al. 2009). It has been identified that there are two channels of reaction to methanol (i) reaction of reverse water-gas shift by decomposition of CO$_2$ to CO, and (ii) the mechanism of well-regarded by an intermediate of formate (Chen et al. 2003). Because of the presence of route for CO$_2$ splitting, the formation of CO cannot be avoided, the barrier of hydrogenation being quite comparable to the adsorption energy of CO. even though the surface species of H$_2$COO and HCOO can be identified, they are still not the intermediate key to methanol. The route of hydrolysis plays important roles in the formation of methanol and also removal of O. low rate of conversion for fixation of CO$_2$ can be related to the lack of active sites for reaction or adsorption of CO$_2$, where the linkage of (Zr)$_2$OCu interface is required.

In order to increase the selectivity and also enhance the conversion rate of CO$_2$, the property of the interface must be optimised by controlling affinity of O at the oxide cationic site. Recently, it has been pointed out by a study that formation of CH$_3$OH does not occur by the intermediate of formate (Yong et al. 2010). The hydrogenation of adsorbed species of formate on the Cu/SiO$_2$ have been studied by these authors. From the results, it has been indicated that the synthesis of methanol on Cu will not results in direct hydrogenation of (bidentate) formate species in steps that is simple and involve the adsorbed of H species alone. Furthermore, same results were given by experiments that were performed on both unsupported copper and supported (Cu/SiO_2) catalyst. It shows that only the chemistry of metal surface was involved in the reaction mechanism of the methanol synthesis.

The adlayer of bidentate formate pre-exposure to oxidation by N_2O or O_2 results in a change to a configuration of monodentate. Production of methanol in significant quantities can occurs by titration of this coadsorbed layer of monodentate formate/O in dry hydrogen, even though the formate decomposition to H_2 and CO_2 stays the dominant pathway of reaction. During this reaction, it can be observed that the water will be produced simultaneously as the surface of the copper is also reduced. From these results, it can be indicated that co-adsorbates which relates to species of water-derived or surface oxygen can be critical to the production of methanol on Cu, and maybe the hydrogenation of adsorbed formate to adsorbed methoxyl will be assisted. New perspective has been given by these studies to the issue regarding the direct synthesis of methanol from CO_2. In fact, up to now, most of the catalysts that has been utilised were essentially based on the active components coupling in the reaction of reverse water-gas shift (RWGS),

$$CO_2 + H_2 \rightarrow CO + H_2O \qquad (11.7)$$

to methanol from syngas

$$CO + 2H_2 \rightarrow CH_3OH \qquad (11.8)$$

When the fraction of CO_2 in the feed of syngas is small, rapidly the reaction will be proceeds to the equilibrium composition. When the CO_2 concentration in the feed was reduced by a factor of 10, the reaction will be much more sluggish, showing the importance of carbon dioxide as a precursor to methanol. In the presence of steam, initially, the reaction of WGS between CO and H_2O is rapidly forming CO_2 and H_2. With a ratio of $H_2:CO_2$ is 3:1 feed the reaction of RWGS, initial rapid conversion of CO_2 and H_2 to H_2O and CO will occurs, but the methanol conversion will be very slow due to the amount of remains CO_2 which is not converted are large. This is results from the water which act as inhibitor of the catalyst.

Water is not an issue when syngas is used as feed because in the reaction of WGS, it will be consumed rapidly. While, when H_2/CO_2 is used as feed, it is generated from the RWGS, determining the described effect of inhibition. CO and CO_2 will coexist in an amount which depends on the equilibria of thermodynamic, when the catalyst is active in both synthesis of methanol and RWGS reactions. The aspects which are critical in governing the behaviour of the catalytic is the site coverage by strongly chemisorbed CO_2 and the effects of the water formed (Eq. 11.7) on the reactivity of Eq. (11.8). However, a direct path on conversion of CO_2 to methanol shows the possibility to design a new type of catalysts which is not active in the reaction of RWGS, therefore the problem of lower productivity in synthesis of methanol using CO_2 instead of a feed based on CO can be eliminated.

There is therefore space to design new catalysts which are specific for direct conversion of CO_2 to methanol, where this route will be more attractive economically. To optimise the properties of the catalyst, there are new approaches which are possible to be carried out based on the oxides of Cu-Zn consists of various modifiers or dopants, in addition to alumina (Ga$_2$O$_3$, ZrO$_2$, and SiO$_2$) (Liu et al. 2003; Yang et al. 2008; Lim et al. 2009; Ma et al. 2009; Raudaskoski et al. 2009). Conceptually, few other different catalysts have been investigated, but recently, a study on synthesis of methanol from CO$_2$ and H$_2$ on a cluster of Mo$_6$S$_8$ have been suggested as materials that can be developed (Liu et al. 2009). Pd–ZnO on carbon nanotubes also has been considered as a new type of catalyst (Liang et al. 2009; Zhang et al. 2010a). The ability to adsorb a large amount of hydrogen reversible is the peculiarity of carbon nanotube. Due to the presence of complex equilibria in the system of CO/CO$_2$/H$_2$O/CH$_3$OH, the performances of catalyst reported from different data of literature is difficult to be compared. This is because, no comparison can be done with different compositions of feed.

One of the most detailed studies have been reported by Saito and Murata (2004). Evidently, with respect to CO, stronger H$_2$O inhibition effect on the methanol synthesis can be observed using 1:3 ratio of CO$_2$:H$_2$ as feed over a catalyst of Cu/ZnO/ZrO$_2$. Investigation on different oxide supports such as Ga$_2$O$_3$, Al$_2$O$_3$, Cr$_2$O$_3$ and ZrO$_2$ were carried out, and Ga$_2$O$_3$ shows the best productivity. Improvement in the specific activity for synthesis of methanol was allowed by Ga$_2$O$_3$, while the stability and dispersion of particles can be improved by Al$_2$O$_3$, SiO$_2$ or ZrO$_2$. Better performances were shown by the systems of multicomponent, (Cu/ZnO/ZrO$_2$/Al$_2$O$_3$/SiO$_2$) with stable productivity of about 600 g of CH$_3$OH/l cat at temperature of 250 °C under realistic reaction conditions (Saito and Murata 2004; Toyir et al. 2009).

For comparison, using the same catalyst under similar conditions with higher ratio of CO/CO$_2$, for instance, a feed of CO (25%)/CO$_2$ (6%)/H$_2$ (69%), the productivity of methanol is three times higher. It has been evidenced that, in order to match the performances using syngas, the performances of the catalyst still need to be improved. Due to the strong inhibition of water, a new process to convert CO$_2$ to methanol must include a continuous water removal from the reactors. Catalytic distillation and also the utilisation of inorganic water membranes of perm selective are the possible solutions. Even though there are no specific results have been reported in any literature, the latter solution is preferred. Membranes which are recoverable and also suitable are those based on the films of hydrophilic nanopore zeolite (NaA) over a support of ceramic tubular and was developed for water ethanol solutions pervaporation (Kazemimoghadam and Mohammadi 2010).

11.7.2   DME from CO$_2$

DME is an alternative fuel which is economical and also clean (high cetane number, low boiling point). Recently, there are several developments of technology for selective olefins production, especially propylene either from DME or from methanol. The development of technology for olefins conversion from methanol in which DME is generated as an intermediate are fairly advanced in Lurgi and UOP. While, the R&D on conversion of DME to olefins was interested by JGC and Idemitsu in Japan. Another reaction that is valuable is the carbonylation of DME to methyl acetate. The synthesis of DME from H_2/CO_2 must be considered as an extension for synthesis of methanol to shift the equilibrium by utilising hybrid catalysts.

$$2CH_3OH \leftrightarrow CH_3OCH_3 + H_2O \tag{11.9}$$

Catalysts utilised are thus a catalyst combination for synthesis of methanol from H_2/CO_2 and acid catalyst, usually zeolite. About 55% of DME selectivity and 17% of DME yields can be observed when the Y-zeolite was combined with $Cu/ZnO/Al_2O_3$ catalyst (Arakawa 1998). While, about 70% of DME selectivity can be observed when SAPO-34 zeolite was used as catalyst. Improvement in the performances can be achieved by using zirconia as the support for Cu/ZnO catalyst and also by tuning the zeolite acidity (Ihm et al. 2003). The performances also can be improved by structured reactor with a configuration of dual-bed catalyst. It has been observed that the formation of DME also can be favoured by addition of Pd to a hybrid $CuO-ZnO-Al_2O_3-ZrO_2/HZSM-5$ catalyst. It has been found that the presence of Al_2O_3 also was also beneficial for DME (Wang and Zeng 2005), even if the maximum DME selectivity was 40%. Recently, some catalysts which are bifunctional have been reported and showing some remarkable improvements: $CuO-TiO_2-ZrO_2/HZSM-5$ (Wang et al. 2009), $CuO-ZnO-Al_2O_3-ZrO_2/HZSM-5$ (An et al. 2008), and also $CuO-ZnO-Al_2O_3/HZSM-5$ (Zhao et al. 2007).

11.8 Conclusions

When stages of fermentation are present, biorefineries will give relevant impact on the emissions of greenhouse gas. The requirement to reduce this impact, by chemically utilising CO_2 that associated to processes of fermentation because of high purity of these emissions have been discussed. To adopt a strategy to decrease the emissions of greenhouse gas and increase the renewable energy share, it has been sown that conversion to chemicals and fuels can be relevant in a synergic approach. Positive impact also has been shown by this approach in terms of offering new opportunities to synthesise chemicals that have high-value-added in order to improve the economics process. The DME and methanol production have been the focused of the discussion, because they are considered to be a choice of product which is preferable. Different stages are required for current technologies and it have to be further developed to be competitive, even though the actual results are not too far to be exploited.

In future, reducing the steps and integrate them in a single device progressively is necessary. The roadmap of indicative has been presented, and it shows that in a medium term, the development of inverse alcohol fuel cells and solar cells is required in order to produce H_2 which is integrated with units of PV with electrolysis sites. In the longer term, devices which are fully integrated is required to capture sunlight for production of electrons and protons from water, and in a dark zone, the electrolytic sites for the methanol reduction. In shorter term, the focus on developing different sequential devices for production of H_2 and then convert CO_2 catalytically using this renewable H_2 must be focalised. Even if formally the reaction between CO_2 and He will give DME or methanol as their products is a reaction that has been established which involves sequential steps, it was shown in a survey that further improvements to this reaction is required. Particularly, the composition changes from syngas to feed of $H_2 + CO_2$ results in new problem related to effects of water inhibition that will affect the productivity as well as the economics of the process. However, it is possible to have a new catalytic pathway to bypass this issue by opening new perspectives. Development of new engineering solutions which is based on water permselective membranes and catalytic distillation is one of the alternatives.

References

Ajhar M, Travesset M, Yüce S, Melin T (2010) Siloxane removal from landfill and digester gas–a technology overview. Biores Technol 101(9):2913–2923

An X, Zuo Y-Z, Zhang Q, D-z Wang, Wang J-F (2008) Dimethyl ether synthesis from CO_2 hydrogenation on a CuO–ZnO–Al$_2$O$_3$–ZrO$_2$/HZSM-5 bifunctional catalyst. Ind Eng Chem Res 47(17):6547–6554

Appels L, Baeyens J, Degrève J, Dewil R (2008) Principles and potential of the anaerobic digestion of waste-activated sludge. Prog Energy Combust Sci 34(6):755–781

Arakawa H (1998) Research and development on new synthetic routes for basic chemicals by catalytic hydrogenation of CO_2. Stud Surf Sci Catal 114:19–30

Boyer L, Vega J, Klasson K, Clausen E, Gaddy J (1992) The effects of furfural on ethanol production by saccharomyces cereyisiae in batch culture. Biomass Bioenerg 3(1):41–48

Yong Y, Mims CA, Disselkamp RS, Ja-Hun K, Peden CHF, Campbell C (2010) (Non)formation of methanol by direct hydrogenation of formate on copper catalysts. J Phys Chem 100 (114):17205–17211

Centi G, Perathoner S (2009) Opportunities and prospects in the chemical recycling of carbon dioxide to fuels. Catal Today 148(3):191–205

Centi G, Perathoner S (2010) Towards solar fuels from water and CO_2. Chemsuschem 3(2):195–208

Centi G, Perathoner S (2011) CO_2-based energy vectors for the storage of solar energy. Greenhouse Gases Sci Technol 1(1):21–35

Centi G, Lanzafame P, Perathoner S (2011a) Analysis of the alternative routes in the catalytic transformation of lignocellulosic materials. Catal Today 167(1):14–30

Centi G, Perathoner S, Passalacqua R, Ampelli C (2011b) Solar production of fuels from water and CO_2. In: Nazim Z, Muradov T, Nejat V (eds) Carbon-neutral fuels and energy carriers series: green chemistry and chemical engineering CRC Press (Taylor & Francis group), Boca Raton, FL (US), pp 291–323

Centi G, Quadrelli EA, Perathoner S (2013) Catalysis for CO_2 conversion: a key technology for rapid introduction of renewable energy in the value chain of chemical industries. Energy Environ Sci 6(6):1711–1731

Chang J, Fu Y, Luo Z (2012) Experimental study for dimethyl ether production from biomass gasification and simulation on dimethyl ether production. Biomass Bioenerg 39:67–72

Chen YX, Miki A, Ye S, Sakai H, Osawa M (2003) Formate, an active intermediate for direct oxidation of methanol on Pt electrode. J Am Chem Soc 125(13):3680–3681

Cherubini F, Jungmeier G (2010) LCA of a biorefinery concept producing bioethanol, bioenergy, and chemicals from switchgrass. Int J Life Cycle Assess 15(1):53–66

Cherubini F, Strømman AH (2011) Chemicals from lignocellulosic biomass: opportunities, perspectives, and potential of biorefinery systems. Biofuels Bioprod Biorefin 5(5):548–561

Clausen LR, Elmegaard B, Houbak N (2010a) Technoeconomic analysis of a low CO_2 emission dimethyl ether (DME) plant based on gasification of torrefied biomass. Energy 35(12):4831–4842

Clausen LR, Houbak N, Elmegaard B (2010b) Technoeconomic analysis of a methanol plant based on gasification of biomass and electrolysis of water. Energy 35(5):2338–2347

da Silva AL, Dick LFP, Müller IL (2012) Performance of a PEMFC system integrated with a biogas chemical looping reforming processor: a theoretical analysis and comparison with other fuel processors (steam reforming, partial oxidation and auto-thermal reforming). Int J Hydrogen Energy 37(8):6580–6600

De Wild P, Nyqvist R, De Bruijn F, Stobbe E (2006) Removal of sulphur-containing odorants from fuel gases for fuel cell-based combined heat and power applications. J Power Sources 159 (2):995–1004

Demirbas A (2008) Biofuels sources, biofuel policy, biofuel economy and global biofuel projections. Energy Convers Manag 49(8):2106–2116

Dewil R, Appels L, Baeyens J (2006) Energy use of biogas hampered by the presence of siloxanes. Energy Convers Manag 47(13):1711–1722

Ihm S-K, Baek S-W, Park Y-K, Jeon J-K (2003) CO_2 hydrogenation over copper-based hybrid catalysts for the synthesis of oxygenates. ACS Publications

Jiang Z, Xiao T, Vá Kuznetsov, Pá Edwards (2010) Turning carbon dioxide into fuel. Philos Trans R Soc Lond A Math Phys Eng Sci 368(1923):3343–3364

Kazemimoghadam M, Mohammadi T (2010) The pilot-scale pervaporation plant using tubular-type module with nano pore zeolite membrane. Desalination 255(1):196–200

Lange JP, van der Heide E, van Buijtenen J, Price R (2012) Furfural—a promising platform for lignocellulosic biofuels. Chemsuschem 5(1):150–166

Lantz E, Wiser R, Hand M (2012) The past and future cost of wind energy. Report No. NREL/TP-6A20-53510. National Renewable Energy Laboratory, Golden, CO,

Lau C, Allen D, Tsolakis A, Golunski SE, Wyszynski M (2012) Biogas upgrade to syngas through thermochemical recovery using exhaust gas reforming. Biomass Bioenerg 40:86–95

Liang X-L, Dong X, Lin G-D, Zhang H-B (2009) Carbon nanotube-supported Pd–ZnO catalyst for hydrogenation of CO_2 to methanol. Appl Catal B 88(3):315–322

Lim H-W, Park M-J, Kang S-H, Chae H-J, Bae JW, Jun K-W (2009) Modeling of the kinetics for methanol synthesis using $Cu/ZnO/Al_2O_3/ZrO_2$ catalyst: influence of carbon dioxide during hydrogenation. Ind Eng Chem Res 48(23):10448–10455

Liu X-M, Lu G, Yan Z-F, Beltramini J (2003) Recent advances in catalysts for methanol synthesis via hydrogenation of CO and CO_2. Ind Eng Chem Res 42(25):6518–6530

Liu P, Choi Y, Yang Y, White MG (2009) Methanol synthesis from H_2 and CO_2 on a Mo_6S_8 cluster: a density functional study. J Phy Chem A 114(11):3888–3895

Ma J, Sun N, Zhang X, Zhao N, Xiao F, Wei W, Sun Y (2009) A short review of catalysis for CO_2 conversion. Catal Today 148(3):221–231

Mandl MG (2010) Status of green biorefining in Europe. Biofuels Bioprod Biorefin 4(3):268–274

McBean EA (2008) Siloxanes in biogases from landfills and wastewater digesters. Can J Civ Eng 35(4):431–436

Menegazzo F, Pinna F, Signoretto M, Trevisan V, Boccuzzi F, Chiorino A, Manzoli M (2008) Highly dispersed gold on zirconia: characterization and activity in low-temperature water gas shift tests. Chemsuschem 1(4):320–326

Mota N, Alvarez-Galvan C, Navarro R, Fierro J (2011) Biogas as a source of renewable syngas production: advances and challenges. Biofuels 2(3):325–343

Osorio F, Torres J (2009) Biogas purification from anaerobic digestion in a wastewater treatment plant for biofuel production. Renew Energy 34(10):2164–2171

Piccolo C, Bezzo F (2009) A techno-economic comparison between two technologies for bioethanol production from lignocellulose. Biomass Bioenerg 33(3):478–491

Powell E, Hill G (2009) Economic assessment of an integrated bioethanol–biodiesel–microbial fuel cell facility utilizing yeast and photosynthetic algae. Chem Eng Res Des 87(9):1340–1348

Quadrelli EA, Centi G, Duplan JL, Perathoner S (2011) Carbon dioxide recycling: emerging large-scale technologies with industrial potential. Chemsuschem 4(9):1194–1215

Raudaskoski R, Turpeinen E, Lenkkeri R, Pongrácz E, Keiski R (2009) Catalytic activation of CO_2: use of secondary CO_2 for the production of synthesis gas and for methanol synthesis over copper-based zirconia-containing catalysts. Catal Today 144(3):318–323

Rosenberg JN, Mathias A, Korth K, Betenbaugh MJ, Oyler GA (2011) Microalgal biomass production and carbon dioxide sequestration from an integrated ethanol biorefinery in Iowa: a technical appraisal and economic feasibility evaluation. Biomass Bioenerg 35(9):3865–3876

Ryckebosch E, Drouillon M, Vervaeren H (2011) Techniques for transformation of biogas to biomethane. Biomass Bioenerg 35(5):1633–1645

Saito M, Murata K (2004) Development of high performance Cu/ZnO-based catalysts for methanol synthesis and the water-gas shift reaction. Catal Surv Asia 8(4):285–294

Schleicher-Tappeser R (2012) How renewables will change electricity markets in the next five years. Energy policy 48:64–75

Supekar SD, Skerlos SJ (2014) Market-driven emissions from recovery of carbon dioxide gas. Environ Sci Technol 48(24):14615–14623

Taarning E, Nielsen IS, Egeblad K, Madsen R, Christensen CH (2008) Chemicals from renewables: aerobic oxidation of furfural and hydroxymethylfurfural over gold catalysts. Chemsuschem 1(1–2):75–78

Tang Q-L, Hong Q-J, Liu Z-P (2009) CO_2 fixation into methanol at Cu/ZrO_2 interface from first principles kinetic Monte Carlo. J Catal 263(1):114–122

Toyir J, Miloua R, Elkadri N, Nawdali M, Toufik H, Miloua F, Saito M (2009) Sustainable process for the production of methanol from CO_2 and H_2 using Cu/ZnO-based multicomponent catalyst. Physics Procedia 2(3):1075–1079

Wang J, Zeng C (2005) Al_2O_3 effect on the catalytic activity of $Cu–ZnO–Al_2O_3–SiO_2$ catalysts for dimethyl ether synthesis from CO_2 hydrogenation. J Nat Gas Chem 14(3):156–162

Wang S, Mao D, Guo X, Wu G, Lu G (2009) Dimethyl ether synthesis via CO_2 hydrogenation over $CuO–TiO_2–ZrO_2$/HZSM-5 bifunctional catalysts. Catal Commun 10(10):1367–1370

Wellisch M, Jungmeier G, Karbowski A, Patel MK, Rogulska M (2010) Biorefinery systems—potential contributors to sustainable innovation. Biofuels Bioprod Biorefin 4(3):275–286

Woods J, Black M, Murphy R (2008) Future feedstocks for biofuel systems. Biofuels: environmental consequences and interactions with changing land use. In: Proceedings of the scientific committee on problems of the environment (SCOPE) international biofuels project rapid assessment, pp 22–25

Xu J, Zhou W, Li Z, Wang J, Ma J (2009) Biogas reforming for hydrogen production over nickel and cobalt bimetallic catalysts. Int J Hydrogen Energy 34(16):6646–6654

Xu Y, Isom L, Hanna MA (2010) Adding value to carbon dioxide from ethanol fermentations. Biores Technol 101(10):3311–3319

Yang R, Yu X, Zhang Y, Li W, Tsubaki N (2008) A new method of low-temperature methanol synthesis on $Cu/ZnO/Al_2O_3$ catalysts from $CO/CO_2/H_2$. Fuel 87(4):443–450

Zhang Q, Zuo Y-Z, Han M-H, Wang J-F, Jin Y, Wei F (2010) Long carbon nanotubes intercrossed Cu/Zn/Al/Zr catalyst for CO/CO_2 hydrogenation to methanol/dimethyl ether. Catal Today 150 (1):55–60

Zhao Y, Chen J, Zhang J (2007) Effects of ZrO_2 on the performance of $CuO–ZnO–Al_2O_3$/HZSM-5 catalyst for dimethyl ether synthesis from CO_2 hydrogenation. J Nat Gas Chem 16(4):389–392

Zinoviev S, Müller-Langer F, Das P, Bertero N, Fornasiero P, Kaltschmitt M, Centi G, Miertus S (2010) Next-generation biofuels: survey of emerging technologies and sustainability issues. Chemsuschem 3(10):1106–1133

Printed in the United States
By Bookmasters